Informationsmanagement in Theorie und Praxis

Reihe herausgegeben von
A. Gadatsch, Sankt Augustin, Deutschland
D. Schreiber, Sankt Augustin, Deutschland

Das Informationsmanagement steht im Spannungsfeld zwischen betriebswirtschaftlichen und technischen Herausforderungen. Aktuelle Themen wie Digitalisierung, Big Data und die sich hieraus entwickelnden disruptiven Geschäftsmodelle spiegeln sich in starken Veränderungen wieder, sowohl in der Theorie, als auch in der Praxis. Die Schriftenreihe „Informationsmanagement in Theorie und Praxis" greift diese Themen auf, sowohl in der Forschung in Form herausragender Dissertationen, als auch in der Umsetzung durch exzellente Masterarbeiten.

Information Management is located between business and technical challenges. Current topics such as Digitalization, Big Data and the resulting disruptive business models are reflected in strong changes, both in theory and in practice. The series "Information Management in Theory and Practice" sets a focus on these topics, both in research theses (Ph. D.), as well as in practice theses (MSc).

Weitere Bände in der Reihe http://www.springer.com/series/16088

Daryoush Daniel Vaziri

Facilitating Daily Life Integration of Technologies for Active and Healthy Aging

Understanding Demands of Older Adults in Health Technology Design

Daryoush Daniel Vaziri
Bonn, Germany

Dissertation Universität Siegen, 2018 u.d.T.: Daryoush Daniel Vaziri: „Facilitating Daily Life Integration of Technologies for Active and Healthy Aging. Understanding and Addressing Heterogeneous Demands of Older Adults and Relevant Stakeholders in Health Technology Design"

Dissertation zur Erlangung des Dr. rer. pol. bzw. Dr. iur. der Fakultät III – Wirtschaftswissenschaften, Wirtschaftsinformatik und Wirtschaftsrecht der Universität Siegen.

Finishing year: 2018
Faculty director: Prof. Dr. Volker Wulf
First reviewer: Prof. Dr. Volker Wulf
Second reviewer: Prof. Dr. Dirk Schreiber
Third reviewer: Prof. Dr. Gerd Morgenthaler
Day of the oral examination: 15 February 2018

ISSN 2524-4205   ISSN 2524-4213   (electronic)
Informationsmanagement in Theorie und Praxis
ISBN 978-3-658-22874-3   ISBN 978-3-658-22875-0   (eBook)
https://doi.org/10.1007/978-3-658-22875-0

Library of Congress Control Number: 2018950040

Springer Vieweg
© Springer Fachmedien Wiesbaden GmbH, part of Springer Nature 2018

Printed on acid-free paper

This Springer Vieweg imprint is published by the registered company Springer Fachmedien Wiesbaden GmbH part of Springer Nature
The registered company address is: Abraham-Lincoln-Str. 46, 65189 Wiesbaden, Germany

For my parents

Acknowledgements

Personal and professional development are key elements in the long process of becoming a Ph.D. Along this process a number of people are involved and accompany the Ph.D. student in their undertaking. In my case, this journey began with Prof. Dr. Reiner Clement, Prof. Dr. Andreas Gadatsch and Prof. Dr. Dirk Schreiber, all former professors of mine at University of Bonn-Rhein-Sieg. I thank Reiner Clement, who I worked for as a student assistant during my studies. Back then, he already anticipated the path I would take a couple of years later, by remarking my scientific capabilities and motivating me to become a doctor. I would have been proud to tell him in person that I succeeded. I thank Andreas Gadatsch, who accompanied me when I made my first steps in the scientific community. He supported me in scientific writing and enabled my first scientific experiences. I deeply thank Dirk Schreiber for bringing me back to University after I finished my Bachelor degree and started to work at Deutsche Telekom. It was he, who prepared the road for me to become a doctor, and I am grateful for his tireless effort, his feedback and support during my study and his trust he put in me from the beginning.

On my journey, I had the pleasure to get to know Prof. Dr. Volker Wulf and Dr. Rainer Wieching from the University of Siegen. Volker Wulf did not know me well in the beginning, but still took a chance with me and agreed to supervise me in cooperation with Dirk Schreiber. I thank him deeply for the leap of faith, his guidance and feedback during my study. Special thanks goes to Rainer Wieching for his support during the last years. I am grateful for the many discussions and conversations that helped me to improve my research skills. I cannot thank him enough for his valuable feedback and input during the finishing phase.

I thank Konstantin Aal, Corinna Ogonowski and David Unbehaun for their cooperation and collaboration in the studies over the last years.

Finally yet importantly, I thank my family and especially my wife who were there for me at times I struggled or lost faith. Their moral support, encouragement and patience made this dissertation possible.

List of figures

List of tables

Abbreviations

Abstract

Diseases and related health problems in the ageing society increase dependency of older adults and their need for care. A consequence in most cases is reduced quality of life of older adults. On top of that, healthcare systems have to face considerable financial burdens in ageing societies. Hence, there is a large interest in innovative and effective ICT-based solutions that delay older adults' need for healthcare services, improve their quality of life and thus mitigate the pressure on healthcare systems. Here, a major challenge is the design of technologies that support active and healthy ageing (AHA) and allow a seamless integration into older adults' daily lives in order to create opportunities for long-term use and sustainable health impacts.

This thesis will present results of a participatory design process for technologies that support AHA in older adults. The design process involved older adults and relevant secondary stakeholders in the respective healthcare system. In a mixed methods approach, qualitative and quantitative methods were combined to create a detailed understanding of older adults' practices and attitudes with respect to health, quality of life and technology use for AHA support and their interactions with relevant stakeholders. Qualitative and quantitative data was collected from older adults aged 60+ and relevant secondary stakeholders like caregivers, policy makers or health insurance companies. Study setups were embedded in living lab structures and randomised controlled trial (RCT) studies. Results illustrate that the combination of multiple AHA technologies and their alignment towards older adults' social contexts and environments may increase their motivation to use technologies for AHA support and address their heterogeneous practices and attitudes more accurately. Furthermore, findings show that perspectives on health, quality of life and AHA technology use of older adults and relevant stakeholders considerably differ and affect their use of AHA technologies. Contradictive perspectives refer to different conceptions of independence, well-being, trust and privacy, which need to be negotiated for the design of technologies for AHA support to facilitate daily life integration.

Based on these findings, the author designed a prototype for an ICT-based solution that combines and aligns multiple AHA technologies and supports collaboration and cooperation between older adults and relevant stakeholders with respect to AHA. The evaluation of the prototype illustrated that technologies for AHA support that consider different health areas and involve older adults' social environment may facilitate the integration into daily lives of older adults and thus create opportunities for long-term use and sustainable health impacts. Finally, the applied methodological approach in this thesis may support sustainable implementation of technologies for AHA support into healthcare systems by easing the uptake of re-search results for primary end-users and relevant secondary stakeholders.

1 Introduction

1.1 Motivation

Worldwide, societies transit towards what is called an ageing population. The proportion of people aged 60 years and above steadily increases, while birth rates stagnate or even decline (Active Ageing, 2002; Barry et al., 2015; Centers for Disease Control and Prevention, 2013). As a consequence, healthcare costs such as stays at retirement homes or caregiving increase dramatically. This development puts considerable pressure on socio-economic systems of affected countries and calls for counteractions. In this context, policy makers and healthcare professionals have an increasing interest in solutions that can help delay the need for healthcare services, prolong independent living, and support older people in what is now termed active healthy ageing (AHA) (Publications Office of the European Union, 2012; Rechel et al., 2013; Vines et al., 2015). Therefore, governments like the European Commission frequently fund programs and projects with respect to AHA (European Commission, 2013). The development of innovative information and communication technology with a focus on lifestyle change and health improvement, is a key aspect in many of those initiatives. AHA technologies like activity monitors, digital dietary and exercise coaches or virtual exercise environments aim at easing the access to prevention programs, while at the same time supporting older adults to pursue a healthy lifestyle. Continuous long-term use of AHA technologies and prevention programs are crucial aspects to achieve health improvements in older adults (Uzor and Baillie, 2014). Here, literature does not provide a clear definition of what is called "long-term use" in the context of health technology design. Previous research though suggests that studies should at least span several months to investigate such long-term use and effects on older adults' health and quality of life (Gerling et al., 2015; Müller et al., 2015a; Ogonowski et al., 2016; Vaziri et al., 2017; Wan et al., 2014). However, investigating long-term use of AHA technologies by older adults and possible effects on their health might require observations over several years to yield reliable data. One important underlying factor for long-term use is integration of such technologies into older adults' daily lives (Hänsel et al., 2015; Schlomann et al., 2016). Hence investigating the integration of AHA technologies into older adults' daily lives may support researchers to draw more reliable conclusions about long-term use and effects of technologies for AHA support. Nevertheless, the design of AHA technologies is a complex endeavor, as heterogeneity of practices and attitudes in the population of older adults is strong. In the field of AHA technology design, researchers make use of qualitative and quantitative methods to evaluate the efficacy of health technologies or factors like usability, user experience or user acceptance, which are

© Springer Fachmedien Wiesbaden GmbH, part of Springer Nature 2018
D. D. Vaziri, *Facilitating Daily Life Integration of Technologies for Active and Healthy Aging*, Informationsmanagement in Theorie und Praxis, https://doi.org/10.1007/978-3-658-22875-0_1

important to facilitate daily life integration of such technologies by older adults. However, in most cases, research rather specializes in qualitative or quantitative methods, instead of aiming for converging both methods (Brannen, 2005). The author of this thesis argues, it is debatable to what extent the sole use of qualitative or quantitative methods is capable to address heterogeneous practices, attitudes and perspectives of older adults and relevant stakeholders like doctors, health insurance companies or policy makers in health technology design for facilitating integration into daily lives of older adults. Without question though, designers need to be aware of the fact that the use of AHA technologies considerably affects all aspects of older adults' social lives (Pipek and Wulf, 2009). Therefore, the design of information and communication technology (ICT) artefacts in the context of technologies for AHA support needs to consider and understand the social arrangements of older adults who apply these artefacts. Gaining such understandings, moreover, necessitates approaches which place designers and users in equal positions (Rohde et al., 2009). However, achieving equality in the design process turns out to be a challenging process. The literature mentions several challenges in the design of AHA technologies for older adults, for instance, (1) older adults' limited capability to understand technical terms or artefacts and articulate requirements and obligations, (2) their longer learning curve and need for multiple iterations to get used to AHA technologies and (3) their need for social support infrastructures not only for technical issues but for social participation (Chaudhry et al., 2016; Eisma et al., 2003; Lindsay et al., 2012). To address such challenges, research has indicated the need for qualitative methodologies that support a better understanding of technological requirements for the design of technologies appropriate for older adults (Keith and Whitney, 1998; Lindsay et al., 2012). However, adoption and acceptance of health technologies by older adults is substantially influenced by interactions and relations with their social environment and therefore relevant stakeholders within the healthcare system play a major part and deserve critical attention in health technology design (Brannen, 2005; Green et al., 2015; Hammersley, 2000). On the one side, this calls for an active collaboration and cooperation in the design process between older adults, researchers and industry (Newell et al., 2006). Here, participatory design (PD) constitutes an appropriate framework to identify and address design challenges in work with older adults and provides for a methodological eclecticism to support a better understanding of older adults' specific needs and their interactions and relations with relevant stakeholders (Grönvall and Kyng, 2013; Lindsay et al., 2012; Siek et al., 2010). On the other side, results of PD research need to address the different requirements and interests of involved stakeholders by providing lucid information and enable relevant secondary stakeholders, like doctors, policy makers or health insurance companies, the uptake of research results (Berwick, 2003; Hammersley, 2000). This becomes particularly important for research embedded in funded research projects that expect the dissemination and sustainable implementation of developed technologies for AHA support and research results (Berwick, 2003; Proctor et al., 2009; Westfall et al., 2007).

The thesis will investigate practices, attitudes and perspectives of older adults with respect to technologies that support AHA and disease prevention in older adults. In this light, AHA technologies will be used as a superordinate term for devices like activity or sleep monitors, exergames or digital dietary or exercise coaches. The case studies will build a detailed understanding of AHA practices and attitudes in older adults across European and Australian populations and illustrate the design of a health technology prototype in a participatory design (PD) process. Furthermore, the case studies will provide insights on interactions and relations between older adults and their social environment, which affect their practices and attitudes with respect to AHA technology use. Finally, this thesis will illustrate how the use of a mixed methods design may support the design of technologies for AHA support and promote sustainable implementation of such technologies into the respective healthcare system. Following, the structure of this thesis will be presented.

1.2 Structure of the thesis

This thesis is structured into three main sections, (1) state of the art, (2) main findings, and (3) reflection and discussion. Chapter 2 will provide the state of the art in health technology research, including the initial motivation for research in this field from a political point of view and proceeding to AHA technologies and the challenges in designing them with and for older adults. Chapter 3 will introduce the research questions for this thesis. Chapter 4 outlines the research design by describing the research settings and the applied methodologies. Chapters 5 to 9 will present the main findings related to the aforementioned research questions. The case studies within these chapters are structured in three main categories: System usage aspects of AHA technologies, perspectives on health, quality of life and technology use of older adults and relevant stakeholders, and a prototype design for an AHA technology for older adults.

System usage aspects of AHA technologies

Chapter 5: Exploring user acceptance and user experience. This chapter evaluates a prototype fall prevention system for older adults with respect to user acceptance and user experience. The prototype was designed in close collaboration and cooperation with end users in the context of an EU research project over three years. The goal of the evaluation was to measure the extent to which the prototype system fulfilled requirements of the target group and to identify remaining challenges for further design of AHA technologies. Therefore, measures for user acceptance, usability and user experience has been applied with participants who used the system over the course of 6 months.

Chapter 6: Analysing effects and usage indicators. This chapter analyses the effects of a fall prevention prototype system used in a randomized controlled trial (RCT) study over a period of 4 months. The main goals were to find out how older adults' system usage affected their practices and attitudes with respect to AHA technology use and their health. Statistical analyses investigated system effectivity and system usage. Based on these findings, further qualitative analyses with participants from a 6 months living lab explained the effects of system usage on older adults pratices and attitudes in more detail and suggested indicators influencing the use of such systems by older adults.

Perspectives on health, quality of life and technology use of older adults and relevant stakeholders

Chapter 7: Exploring health technology user behaviour. This chapter conducts an investigation into predictors for AHA technology use by older adults. Based on quantitative data collected from older adults who previously bought health technologies, statitsical analyses revealed a set of possible predictors for AHA technology use in that target group. In addition, qualitative interviews were conducted with participants from a living lab to understand practices and attitudes that may exemplify and provide meaning to these predictors.

Chapter 8: Engaging Disparate Stakeholder Demands. This chapter deals with disparate perspectives and conceptions of older adults and secondary stakeholders like doctors, policy makers or health insurance companies in the healthcare system. Here, the goal was to provide meaningful insights into relations of older adults and secondary stakeholders and how these influence older adults' willingness to engage with healthy activities and AHA technologies. Therefore, exploratory qualitative interviews were conducted with secondary stakeholders and participants who tested a range of AHA technologies over the course of two months.

Prototype design for an AHA technology for older adults

Chapter 9: Prototype design for a health technology. As a consequence of results and implications made in the previous chapters, this chapter deals with the design of a technology prototype for AHA support that aims at addressing the heterogeneous practices and attitudes of older adults and relevant stakeholders. By using key elements of design case studies, this chapter documents how design decisions were made and how they were addressed in a high-fidelity prototype. The prototype was evaluated with the target group and implications for the future design were derived.

Chapter 10 will summarize the findings of the presented case studies by discussing them in the light of the research questions, which will be presented in chapter 3 of this thesis. The case studies provide a detailed understanding of practices, attitudes and perspectives of older adults and relevant stakeholders with respect to health, quality of life and AHA technology use. The investigation throughout the case studies followed a mixed methods approach and led to a prototype design that addresses heterogeneous practices and attitudes of older adults. The use of mixed methods in the design of AHA technologies for older adults will be discussed. The thesis concludes with suggested implications for future design of technologies for AHA support that may help to address heterogeneity in older adults and the development of AHA technologies that facilitate technology integration into older adults' daily lives and thus promote long-term technology use and sustainable health impacts.

2 State of the art

2.1 Policies and ageing society

The European commission regularly announces research calls, under the name of Active and Healthy Ageing (AHA), to tackle the challenges of an ageing society with innovative health care products and services for older people. Increasingly, projects focus the development of ICT-based interventions for the support of AHA in older people. Such projects are expected to yield solutions that promote sustainability in AHA in the form of long-term use, integration into daily routines and effectiveness in improving health. In this context, ICT-based solutions for AHA must address a shift in the power relationships within healthcare systems, away from the unrestrained authority of the medical professional and towards a more collaborative partnership with patients taking on a greater responsibility and a more active role in managing their own wellbeing. To manage such responsibilities, users need not only to understand the possibilities of AHA technologies but they also need to feel that they have control over how they interact with them (Publications Office of the European Union, 2012). A better understanding of the possibilities that modern health technologies provide in terms of AHA requires education in health literacy (Sørensen et al., 2012). From a policy perspective, a high level of health literacy is considered of substantial benefit for maintaining health (Sørensen et al., 2012; Wolf et al., 2005). At the same time, the empowerment of older people with control over technology and their own health data is required to address the desire to be treated as equals rather than passively monitored subjects (Mihailidis et al., 2008; Steele et al., 2009).

    Despite significant efforts on the part of policy and research in the development of AHA technologies, we often see that in real world contexts primary users lose motivation quickly and stop using these technologies. This happens most often because new technologies for AHA support fail to successfully integrate into daily routines of older people, including interactions with other stakeholders (Ogonowski et al., 2016; Wan et al., 2016). Policy and research might need to think broader and start to consider the healthcare system as a space where all stakeholders, not only primary end users like older people, but also stakeholders like physicians, health insurance companies or policy makers interact and influence one another and affect the sustainability and long-term use of AHA technologies.

© Springer Fachmedien Wiesbaden GmbH, part of Springer Nature 2018
D. D. Vaziri, *Facilitating Daily Life Integration of Technologies for Active and Healthy Aging*, Informationsmanagement in Theorie und Praxis, https://doi.org/10.1007/978-3-658-22875-0_2

2.2 Health monitoring and quantify yourself in older adults

The idea of monitoring technologies is relatively new. In 2004, Dishman discussed the potentials of such technologies to support interventions by collecting data on behaviors and detecting problems in a timely manner (Dishman, 2004). Consequently, a vast amount of research in the health domain concentrated on exploiting those potentials by addressing challenges associated with ageing; for instance physical activity, nutritional or cognitive behavior. Today, monitoring technology applied to self-tracking behavior provides the performance and cost-effectiveness to be distributed among a wide user base. Self-tracking devices support the self-management of a variety of life aspects like sleep, nutrition, exercise or mood through the provision of feedback, made possible by recording and analyzing personal health data related to those areas. In general, the provided feedback follows a persuasive strategy with the goal to help users to change their behavior towards a desirable healthy lifestyle (Fogg, 2007). Examples for such devices in the research context are BeWell (Lane et al., 2014), BiFit (Consolvo et al., 2008) and Fish 'n' Steps (Lin et al., 2006). Additionally, commercial applications like Nike+ or FitBit increasingly enter the market. Many of these technologies have been developed for general populations and not older adults in particular. Nonetheless, much research has been conducted in the space of health applications targeted specifically at older adults. Accordingly, developed systems aim to support functional abilities (Lee and Dey, 2011), physical (Doyle et al., 2010a; Uzor and Baillie, 2013), social (Doyle et al., 2010b) or cognitive (Jimison et al., 2010) well-being. Typically, wearable sensors such as pedometers, blood pressure cuffs or pulse monitors are applied for data collection. However, long-term usage of systems targeting older adults in real environments and older adults' willingness to buy such systems seems limited (Brodaty et al., 2005; Robinson et al., 2009; Wan et al., 2016). Literature suggests that usability and user experience aspects, as well as reliable information channels play a major role in uptake and long-term usage of health-related technologies by older adults (Uzor and Baillie, 2013; Wan et al., 2016). But what seems to be more important when addressing needs of older adults is that health technologies follow a holistic approach, which includes all relevant aspects of health and well-being, rather than focusing on only one (Thompson et al., 2011). Hence, there is a demand for platforms integrating different devices in order to process health-related data, and orchestrate relevant activities, according to the needs and demands of older adults.

2.3 Challenges for the design of health technologies for older adults

Technology can be a very valuable tool for addressing the rising health needs of the aging population. Research on the use of ICT has shown that it may positively impact quality of life for older adults (Jacqueline K. Eastman and Rajesh Iyer,

2004; Schulz et al., 2015) by improving social support and psycho-social well-being (Charness and Schaie, 2003; White et al., 2002). In this context, technological progress appears to improve infrastructures and facilitates connections with the outside world helping older adults avoid or reduce feelings of social isolation and loneliness, for instance by engaging in physical activity with peers over distance (Bradley and Poppen, 2003; Mueller et al., 2007; White et al., 1999; Wulf et al., 2004).

In addition, wearable sensors, mobile devices and exergames have also been used in order to detect early-risk and assess frailty in different domains of ageing, like physical activity, cognition, emotional state and social connectedness and have provided preventive measures for such domains (Botella et al., 2012; Gschwind et al., 2015; Schoene et al., 2015). There is strong evidence for the effectiveness of technology-based interventions for the promotion of health (Fanning et al., 2012; Gschwind et al., 2015; Vaziri et al., 2017). However, a remaining challenge is long-term motivation in older adults to use such technologies. Here, the design of health technologies needs to address heterogeneous requirements and capabilities of older adults, for instance, (1) older adults' limited capability to understand technical terms or artefacts and articulate requirements and obligations, (2) their longer learning curve and need for multiple iterations to get used to health technologies and (3) their need for social support infrastructures not only for technical issues but for social participation (Chaudhry et al., 2016; Eisma et al., 2003; Lindsay et al., 2012). Moreover, privacy and trust play an important role as many current technical solutions collect personal health data in order to provide new opportunities for connecting end users with secondary stakeholders such as doctors, business companies, health insurance companies or the government (Braun, 2013; Heart and Kalderon, 2013; Lee et al., 2013; Miller and Bell, 2012; Morris and Venkatesh, 2000). Health technologies need to address such considerations and should be personalized and tailored to the needs of the end users (Andrews and Williams, 2014; Kiosses et al., 2011). However, the design of health technologies should not only align to the needs of primary end users but must also take into account the perspectives of other relevant stakeholders within the healthcare system, such as physicians, policy makers or health insurance companies. Such challenges require appropriate methodologies and instruments that allow and support collaboration and cooperation of all relevant stakeholders as well as mediation and moderation between them by researchers.

2.4 Engaging stakeholders in the design of AHA technologies

In recent years, a considerable amount of research in the area of designing technologies for AHA has been implemented widely. A large part of that research concentrates on primary end users, examining how technology can motivate people to adapt to a healthy lifestyle or how people interact with health technologies and integrate them into their daily routines (Bisafar and Parker, 2016; Cila et al., 2016; Schaefbauer et al., 2015). While researchers have considered other stakeholders extensively, this work has mainly focused on technology use in professional, more clinical or care related, healthcare contexts (Fitzpatrick and Ellingsen, 2013). Further, very little research has considered the integration of primary end users and other stakeholder perspectives for the design of health technologies for private end users, even though it stands to reason that relevant stakeholder groups have divergent interests that might affect technology design and daily life integration by older adults. In this context, Jacobs et al. (2015) compared health information sharing preferences between breast cancer patients, doctors and navigators. They found discrepancies between stakeholder perspectives, such as the hesitation to share emotional issues like loneliness or satisfaction with care (Jacobs et al., 2015). Gerling et al. (2015) investigated long-term use of motion-based video games in care home settings. They found that the integration of caregiver staff is important for the upkeep of technology by older people in this setting. However, perspectives of caregiver staff on such technologies remained unaddressed (Gerling et al., 2015). Latulipe et al. (2015) investigated technology use of a patient portal by low-income patients of health care centers and professional caregivers. From their study, they derived design considerations like providing system use training, audio lexicons of medical reports or compilations of advice and encouragement messages to make the use of the patient portal for low-income older people and caregivers more palatable (Latulipe et al., 2015).

To design health technologies for AHA support that create opportunities for long-term use in older adults, the inclusion and negotiation of all relevant stakeholder perspectives like older adults, doctors, policy makers or health insurance companies is needed. A major challenge when engaging all stakeholders lies in contradictory perspectives and their interactions regarding health and technology use. Addressing all perspectives and demands is impossible. Therefore, designers have to face trade-offs and make decisions how to deal with them. PD provides some solutions to such challenges and offers a framework for collaboration and cooperation among primary end users, researchers and relevant stakeholders.

2.5   Participatory design

Participatory Design (PD) enables different modes and levels of stakeholder participation in the design process. On the one hand, PD distinguishes a normative and emancipatory direction, which is grounded in the design of information systems in the workplace context (Bjerknes and Bratteteig, 1995). Here, an important aspect is the influence of employees they take in the design of information systems they will work with. On the other hand, a more pragmatic and production-oriented perspective requires the involvement of users in the design process as a fundamental approach to designing meaningful and appropriate products or services. A major challenge here is the mediation and moderation of different stakeholder interests and perspectives and to aim for a balance between stakeholder goals and potential contradictions with respect to IT-design (Dorst, 2006). In this context, we may understand participatory design in accordance to following definition: "Participatory design is, as we see it, no longer primarily a professional issue for software developers, but has to be extended to the relationships between different user-designers, and, beyond that, between them and their clients/customers/ service-seeking citizens in general." (Dittrich et al., 2002). Referring to Bodker et al., participatory design may be seen as a process of mutual experiences and learning among different stakeholders including all relevant stakeholders and designers (Bodker et al., 2004). Here, participation means to involve these users and thus reach a common understanding of practices, attitudes and perspectives. Furthermore, this process may create opportunities to discuss contradictive conceptions, boundaries and conditions relevant to the design context. In order to realize participatory design different methods and tools may be applied (Bodker et al., 2010; Muller, 2003; Muller and Kuhn, 1993). Selecting appropriate methods and tools mainly depends on the context of PD. Different approaches here focus on certain situations and the context of use (contextual design), on use and usability aspects (user-centered design), and on the creation of a space where the user may experience design artefacts (experience design). Regardless of the selected approach, the focus of PD is on people engaging in the process as co-designers (Ehn, 2008). However, management, application and the role of the user may differ in each PD approach. Relevant distinguishing characteristics here are for instance, providing access to relevant information, creating opportunities to take independent decisions or providing a space for alternative arrangements with respect to technology and organization (Clement and Van den Besselaar, 1993). Therefore, designers need to make decisions about such aspects in advance, while at the same time allowing adaptations during the PD process. This requires an appropriate environment that facilitates the handling and alignment of complex cooperation and collaboration among designers, end users and different stakeholders.

2.6 Living lab methodology

In order to realize the theoretical foundations of PD in the work with older adults, living labs with a focus on older adults and related stakeholders can provide the required environment for close collaboration and cooperation between end users, researchers and relevant stakeholders. The living lab concept embodies a systematic user co-creation approach that integrates research and innovation processes. The co-creation, exploration, experimentation and evaluation of innovative ideas, scenarios, concepts and related technological artefacts in real life use cases are key elements of living labs (Abowd et al., 2002; Panek et al., 2007). As all relevant stakeholders are involved, potential adoption and acceptance of products and services by primary end users, as well as interactions with relevant stakeholders can be considered concurrently (Pallot, 2009; Schaffers et al., 2011). Important success factors of living labs are (1) the realistic setting in homes of participants, (2) the situation of research in participants' everyday life and (3) the long-term character of conducted field studies running over several months (Budweg et al., 2012; Müller et al., 2015a, 2015b; Wan et al., 2014).

However, in the domain of healthcare innovations little research exists which exploits the potential of living labs and describes the collaboration and cooperation processes with older adults in detail (Lievens et al., 2010; Mulvenna et al., 2011; Ogonowski et al., 2016; Ståhlbröst, 2004). Previous living lab projects reported on the role of researchers, the handling of skill sets and learning needs, the degree of active involvement, and varying duration and commitment of participants that have been positively affected by following the living lab methodology (Carroll and Rosson, 2013; Ley et al., 2015; Ogonowski et al., 2013). Therefore, living labs constitute an appropriate methodology in the context of PD to learn about practices, attitudes and perspectives of older adults and investigate their complex interactions and relations with other relevant stakeholders.

The research presented in this thesis has been conducted in a research environment where living lab methodology and participatory design conflate into what is called PraxLabs. The PraxLabs approach has been developed at University of Siegen in Germany (Müller et al., 2014). Unlike many living lab approaches, PraxLab research is conducted in real households and every-day life contexts of end-user groups, for instance older adults. The duration of PraxLabs generally spans several months to years. PraxLabs may be considered a holistic approach and allow a thorough investigation of the socio-cultural environment of older adults and thus enable researchers a wider perspective on technology adoption by that target group. The every-day proximity in the design process supports the understanding and creation of a common communication sphere among researchers, older adults and relevant stakeholders and by doing so, promote the design of realistic design artefacts that address visions of future user scenarios more accurately (Müller et al., 2014).

3   Research Questions

Health is known to be a multidimensional construct that includes not only physical, but cognitional, nutritional, sleep, emotional and social aspects (Eberst, 1984; Huber et al., 2011). In this context, large parts of AHA technology research try to improve older adults' health and quality of life in these domains (Mason, 2016) and according to several study results, they seem to be successful in doing so (Conn et al., 2009; Gillespie et al., 2012; Gschwind et al., 2015). However, what we can observe is that the uptake and long-term use of AHA technologies by older adults and related stakeholders is low and sustainable health impacts and changes of practices and attitudes with respect to health barely happen (Di Pasquale et al., 2013; Fitzpatrick and Ellingsen, 2013; Jarman, 2014; Ogonowski et al., 2016; Wan et al., 2016). Here, one problem seems to be that health and quality of life underlie individual conceptions, which define the heterogeneity of older adults' practices and attitudes, and uncovering and appropriately addressing those poses a major challenge in the design of AHA technologies. Another problem seems to be that the significance of interactions and relations between older adults and secondary stakeholders like doctors, health insurance companies or policy makers in the healthcare system are neglected in the design of current AHA technologies. However, such individual practices, attitudes, experiences and perspectives of older adults with respect to health, quality of life and technology use, as well as their complex interactions with relevant stakeholders provide relevant implications for the design of AHA technologies for older adults. Understanding and addressing these heterogeneous implications in the design of AHA technologies may facilitate the integration of such technologies into daily lives of older adults and thus create opportunities for long-term use and sustainable health impacts.

This thesis aims at contributing to the scientific discourse by applying a methodological approach that provides a detailed and meaningful understanding of heterogeneous practices and attitudes in the population of older adults with respect to their context of AHA technology use. Based on this understanding, an AHA technology prototype will be designed and evaluated in a participatory design process with older adults. Subsequently, the thesis will discuss how the design of health technologies may address heterogeneous needs of older adults and their complex interactions and relations with relevant stakeholders to facilitate integration of technologies for AHA support into daily lives of older adults. Finally, the author will put the methodological approach into perspective and discuss how the combination of qualitative and quantitative methods may support the design of AHA technologies and furthermore promote sustainable implementation of such technologies in the healthcare system. The case studies presented in this thesis will address following research questions:

© Springer Fachmedien Wiesbaden GmbH, part of Springer Nature 2018
D. D. Vaziri, *Facilitating Daily Life Integration of Technologies for Active and Healthy Aging*, Informationsmanagement in Theorie und Praxis, https://doi.org/10.1007/978-3-658-22875-0_3

How relevant is the consideration of heterogeneity in older adults for the design of AHA technologies for daily life use?
-      What are factors affecting AHA technology use in older adults?

-      What are relevant artefacts for the design of AHA technologies and how do older adults use them in this context?

How does the social environment of older adults and related secondary stakeholders influence their willingness to engage with technologies for AHA support?
-      What are practices and attitudes of older adults with respect to health, quality of life and AHA technology use?

-      How does the social environment of older adults' affect their practices and attitudes in this context?

How can the design of technologies for AHA support address heterogeneous practices and attitudes of older adults to facilitate daily life integration and thus create opportunities for long-term use and sustainable health impacts?
-      What are key aspects in older adults' practices and attitudes that need to be addressed by technologies for AHA support in order to facilitate their integration into older adults' daily lives?

-      How can health technology design address such aspects?

How can mixed methods designs support health technology design and sustainable implementation of AHA technologies in healthcare systems?
-      Does the combination of qualitative and quantitative methods provide a detailed understanding of older adults practices, attitudes and interactions with relevant stakeholders in the context of AHA?

-      What are the benefits of the convergence of qualitative and quantitative methodologies in the context of sustainable implementation of technologies for AHA support in the healthcare system by secondary stakeholders?

The following section will outline the research design. Firstly, the settings in which the case studies were conducted will be described. Secondly, the author provides an introduction into mixed methods approaches that were applied in this thesis. Afterwards, this section will illustrate how data was collected and analyzed. Finally, the author provides an overview of the case studies underlying this thesis.

4 Research design

To answer the research questions introduced above, this thesis will apply a mixed methods approach. While the rationale for all research presented in this thesis is a qualitative practice based approach (Wulf et al., 2011, 2015b), it is accompanied by quantitative descriptive and statistical methods. As previously mentioned the population of older adults is very heterogeneous and the understanding of their attitudes and perspectives with respect to health, quality of life and technology use may be complex. Therefore, approaching this task making use of both qualitative and quantitative methods may allow the investigation of subtle details and also a broader range of relevant cases and thus may draw a more detailed and complete picture of older adults' practices, attitudes, perspectives and relations to their social environment (Brannen, 2005; Green et al., 2015). The case studies presented in this thesis were mainly conducted within the two international research and development projects iStoppFalls and MY-AHA. Both projects were embedded in settings across Europe and Australia.

4.1 Setting

The presented case studies were conducted in the EU funded projects iStoppFalls (Grant Agreement 287361) and MY-AHA (Grant Agreement 689592). The iStoppFalls project was part of the European 7th framework program and ran from 2011 to 2014. Purpose of the project was to design and develop an ICT-based fall prevention system for older adults to use at home. Main goals were to ensure that the system design is appropriate for the use by older adults and to investigate that the system is effective in reducing fall risk in older adults. For the first goal, living labs with older adults were established to examine their practices and attitudes, relevant for the appropriate design of the system for that target group. The living labs ran over a period of 6 months. For the second goal, a 4-month RCT study was conducted. While the living labs took place in Germany (Siegen), study centers for the RCT were located across three different countries, Australia (Sydney), Germany (Cologne) and Spain (Valencia).

      The outcomes of the iStoppFalls project initiated the subsequent project called MY-AHA. The MY-AHA project is part of the European Horizon 2020 program and runs from 2016 to 2020. Projects' study centers are located across the countries Austria, Italy, Germany and Spain. Based on the iStoppFalls results and other research and development (R&D) outcomes, the project aims at developing an integrative health platform that connects various AHA technologies, as well as prevention and intervention programs to tackle multiple areas of older adults' frailty in the domains physical activity, cognition, sleep, nutrition, mood and social connectedness. In this context, iStoppFalls will be part of the MY-AHA

© Springer Fachmedien Wiesbaden GmbH, part of Springer Nature 2018
D. D. Vaziri, *Facilitating Daily Life Integration of Technologies for Active
and Healthy Aging*, Informationsmanagement in Theorie und Praxis,
https://doi.org/10.1007/978-3-658-22875-0_4

system and provides evidence-based fall risk prevention. During a pre-study, living labs with older adults were established in Germany (Siegen) to learn about their practices and attitudes. The pre-study living lab ran over a period of 3 months. In addition, a network of secondary stakeholders from various organizations relevant in the German healthcare system was established in order to gain insights on interactions and relations between older adults and these stakeholders. At a later stage of the project, the efficacy of the MY-AHA system will be assessed in an 18 months RCT study across the countries Austria (Vienna), Italy (Turin), Germany (Cologne) and Spain (Valencia). Simultaneously, an 18 months living lab study in Siegen will provide space for further investigation into practices and attitudes of older adults with respect to their use of the system and their interactions and relations with secondary stakeholders.

4.2   Mixed methods approach

This thesis follows a concurrent mixed methods approach where qualitative and quantitative methods are applied independently of each other for data collection (Teddlie and Tashakkori, 2006). With respect to data analyses and interpretation both, qualitative and quantitative methods were combined. In this context, a prominent approach in literature is the strategy of triangulation, where the choice of methods intends to analyze one social phenomenon from different perspectives. Here, triangulation assumes that data from different methods corroborates one another (Denzin, 1970). However, literature suggests that data collected from different methods may generate a variety of outcomes, and corroboration of results is just one of at least four possibilities (Brannen, 2005; Hammersley, 1996; Morgan, 1998):

- *Corroboration:* Qualitative and quantitative methods yield the same results.
- *Elaboration:* Quantitative findings are exemplified by qualitative analysis.
- *Complementarity:* Qualitative and quantitative results are different but they generate insight when considered together.
- *Contradiction:* Qualitative and quantitative findings conflict.

The findings presented in this thesis are mainly based on corroboration, elaboration and contradiction of results. Convergence of qualitative and quantitative data, in some cases, generated same results and thus strengthened the arguments and design implications. In other cases, the results were contradictive or required further elaboration. This procedure allowed additional perspectives on research objects and revealed unexpected results and implications for design. The following section will describe how the author proceeded in collecting and analyzing the data.

4.3   Overall methodology

4.3.1   Data collection

The sequence of studies in this thesis started in early 2014 as part of iStoppFalls and consisted of several stages. After the completion of iStoppFalls at the end of 2014, the research on daily life integration of technologies for AHA support continued in the subsequent MY-AHA project that started in 2016 and builds upon the previous iStoppFalls project. Figure 1 provides a time line for the research activities, which are described in the following paragraphs.

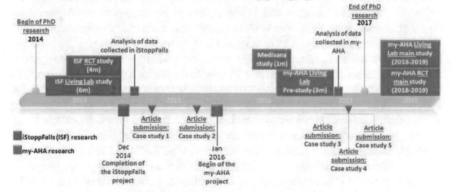

Fig 1. Time line of research activities

First, the author conducted an analysis of effects and usage indicators of an ICT-based fall prevention system called iStoppFalls in order to understand how older adults can benefit from such an ICT-based prevention system, how they use design artefacts and what factors affect their willingness to use the system. Therefore, quantitative data was collected from 153 older adults (Intervention Group=78; control group=75) who participated in a 4 months RCT study across the countries Australia (56), Germany (59) and Spain (36). Data collection was mainly done by continuous data collection of participants' system adherence and their activity, as well as by baseline, mid-term and final assessments that screened participants for characteristics relevant in the context of falls. Section 6.3.2 of this thesis provides a detailed description of this procedure. Further, validated questionnaires with a focus on usability, user experience and user acceptance were distributed to all participants in the RCT. Simultaneously, these questionnaires were distributed among 12 older adults that participated in the living lab study and used the system over a period of 6 months. In order to assess participants' perception of the system's usability, the author applied the system usability scale (SUS) (Brooke, 1996). To learn about participants' enjoyment while exercising with the iStoppFalls system, the author used the physical activity enjoyment scale (PACES) (Kendzierski and DeCarlo, 1991). Finally, to understand underlying factors affecting older adults' technology acceptance, the author designed a questionnaire based on the theoretical foundations of the dynamic model for the re-evaluation of technology (DART) (Amberg et al., 2003). These instruments were applied according to their corresponding instructions described in section 5.2.4 of this thesis. Concurrently to the quantitative data collection, the author conducted interviews with 12 older adults in the living lab study of the project, as well as 28 interviews with participants in the RCT study. All interviews were held and recorded at participants' homes and followed the same semi-structured guideline. During the interviews, participants were mainly asked about their experiences with the iStoppFalls system, their opinion on such systems, their motivation to continue using the iStoppFalls system and their recommendations to improve the system. Afterwards trained research assistants transcribed each interview. In addition, participants in the living labs were visited at their homes to get a better perspective on their daily practices and attitudes with the iStoppFalls system. Interesting events during these observations were documented in field notes. After this stage, the iStoppFalls project was completed and research activities, described in the following paragraphs, were conducted in the context of the MY-AHA project.

Based on the design implications derived from the iStoppFalls project, a second step aimed at exploring older adults' behavior in order to predict their AHA technology use. Here, the goal was to get a better understanding of the underlying factors that influence older adults' use of AHA technologies. Therefore, the author distributed quantitative questionnaires among 203 people who previously bought health technologies. Participants were recruited among customers of Medisana, a company developing and distributing diverse health technology devices like activity or sleep monitors. The questionnaires used validated scales based on the theories of Davis' technology acceptance model (Davis, 1989) and Bandura's social cognitive theory (Bandura, 1986) in order to assess participants' perceptions on technical and social factors with respect to health technology use. A detailed description of the used scales and their application is provided in section 7.2.3 of this thesis. In order to provide a detailed and more complete understanding of the underlying factors, the author conducted additional interviews with 15 older adults that use AHA technologies and take part in a pre-study living lab of the MY-AHA project. Goal of the pre-study living lab is the investigation of older adults' practices and attitudes with respect to AHA technology use. Living lab participants in the pre-study were recruited from local senior organizations. Interviews covered the topics health, quality of life, trust, privacy and technology use. All interviews were performed and recorded at participants' homes. Trained research assistants transcribed each interview.

In a final step of the thesis, the author designed an AHA technology prototype that allows the integration of different health technologies for AHA support and disease prevention in older adults. In a design case study, the author used the outcomes of interviews, observations and interactions with participants from the previous research activities and derived specific design challenges. In addition, home visits, observations and workshops with the 15 living lab participants in the pre-study were conducted to learn more about older adults' requirements with respect to the design of AHA technologies. Finally, the prototype was evaluated with these participants to collect implications for the future design.

4.3.2   Qualitative data analyses

In total, 55 semi-structured interviews with community dwelling older adults aged 60+ years across the five countries Australia, Austria, Germany, Italy and Spain were conducted by the author and trained research assistants. In order to participate in the study, older adults were required to live independently within the community. Interviews in foreign languages, for instance in Spanish or Italian, were translated and transcribed into English language by trained local research assistants, who were fluent in their native and English tongue, in order to consider them for the analysis. Additionally, for the analysis in case study 4, trained research assistants executed 19 semi-structured interviews with relevant secondary stakeholders like doctors, policy makers, caregivers or health insurance companies in the German healthcare system. For the analysis of the interviews, the author oriented on Mayring's qualitative content analysis (Mayring, 2000). Therefore, in collaboration with trained research assistants the author coded a selection of transcripts with respect to the aspects relevant to answer the research questions. For instance, the research team derived codes like "integration of technology into daily life routines", "perspectives on health, "motivation for technology use" or "concerns about trust and data privacy" to tag parts within the data material that covered these aspects. In a first step, each researcher applied this procedure and generated a separate code system. In a second step, the research team discussed each code system and refined and assembled them into a final code system. Finally, all transcripts were analyzed using the coherent code system.

4.3.3   Quantitative data analyses

In total, quantitative data from 215 older adults was collected. Based on the quantitative data collected in the iStoppFalls project, the author first analyzed factors influencing the system usage of older adults. Therefore, collected data from the system usability scale (SUS), the physical activity enjoyment scale (PACES) and the dynamic acceptance model for the re-evaluation of technology (DART) were evaluated, following the evaluation instructions described in corresponding literature (Amberg et al., 2003; Brooke, 1996; Kendzierski and DeCarlo, 1991). All in all, 60 completed units of each questionnaire were returned to the author. Results of these analyzes were illustrated in a descriptive manner using spider charts and illustrative diagrams. Further, the author segmented the RCT study population of 153 older adults into the different subgroups age, gender, information technology (IT)-literacy, education, baseline balance and fallers in order to investigate how the usage of iStoppFalls affected fall risk in these subgroups. Therefore, the author applied repeated ANOVAs and student t-tests in order to quantify and analyze the intervention effects on the subgroup's fall risk, compared to the control group. Furthermore, these analyzes helped to explain the effects of system usage on the reduction of fall risk in older adults. For the analyses, the author used SPSS version 22 (SPSS, Inc., Chicago IL).

From the quantitative questionnaires, distributed among older adults in the context of the subsequent MY-AHA project, 188 completed questionnaires returned. In total, 62 questionnaires were selected for the analysis. The author conducted a linear regression analysis in order to evaluate the value of ease of use, technology experience, usefulness, barriers, social support, activity per week, self-efficacy, physical appearance-related expectations and health-related expectations in the prediction of technology use per week. All predictors were entered into the regression equation in the same step.

4.4   Case studies
The following sections of this thesis will present the published and submitted articles, which report on the results of conducted case studies. Each article will be structured with an introduction, followed by a description of the state of the art, the applied methodologies and the illustration and discussion of results.

System usage aspects of innovative health technologies
-   **Case study 1 (Chapter 5): Exploring user acceptance and user experience.** This chapter has been published as a journal paper in European Review of Ageing and Phyiscal Activity (EURAPA; Impact Factor 2.154), 2016: Vaziri D.D., Aal K., Ogonowski C., Von Rekowski T., Kroll M., Marston H.R., Poveda R., Gschwind Y.J., Delbaere K., Wieching R., Wulf V. Exploring User Experience and Technology Acceptance for a Fall Prevention System: Results from a Randomized Clinical Trial and a Living Lab. In: European Review of Aging and Physical Activity. DOI: 10.1186/s11556-016-0165-z.
-   **Case study 2 (Chapter 6): Analysing effects and usage indicators.** This chapter has been published as a journal paper in Internation Journal of Human-Comupter Studies (IJHCS; Impact Factor 2.863), 2017: Vaziri

D. D., Aal K. , Gschwind, Y.J., Delbaere K., Weibert A., Annegarn J., De Rosario H., Wieching R., Randall D., Wulf V. Analysis of effects and usage indicators for a ICT-based fall prevention system in community dwelling older adults. In: International Journal of Human-Computer Studies. DOI: https://doi.org/10.1016/j.ijhcs.2017.05.004.

Perspectives on health, quality of life and technology use of older adults and relevant stakeholders
-   **Case study 3 (Chapter 7): Exploring health technology user behaviour.** This chapter has been submitted as a journal paper in the Behavior & Information Technology Journal (B&IT; Impact Factor 1.388), 2018: Vaziri D.D., Giannouli E., Frisiello A., Kartinnen N.; Wieching R., Schreiber D., Wulf V. Exploring User Behavior to Predict Influencing Factors for Technology-Supported Physical Activity in Older Adults. In: ACM Transactions on Computer-Human Interactions (under review).
-   **Case study 4 (Chapter 8): Engaging Disparate Stakeholder Demands.** This chapter has been submitted as a journal paper in International Journal of Industrial Ergonomics (IJIE; Impact Factor 1.415), 2018: Vaziri D.D., Unbehaun D., Aal K., Hofheinz S., Shklovski I., Wieching R., Schreiber D., Wulf V. Negotiating Contradictions: Engaging Disparate Stakeholder Demands in Designing for Active and Healthy Aging. In: International Journal of Human-Computer Interaction (under review).

Proposal of a health technology design for older adults
-   **Case study 5 (Chapter 9): Prototype design for a health technology.** This chapter has been submitted as a journal paper in Journal of Human-Computer Interaction (JHCI Impact Factor 3.700), 2018: Vaziri D. D., Anslinger M., Unbehaun D., Wieching R., Randall D., Schreiber D., Wulf V. Prototype Design for an Integrative Health Platform to Support Active and Healthy Ageing in Older Adults. In: Journal of Human-Computer Interaction (under review).

5 Exploring User Experience and Technology Acceptance for a Fall Prevention System: Results from a Randomized Clinical Trial and a Living Lab

Abstract

Background. Falls are common in older adults and can result in serious injuries. The iStoppFalls project developed an information and communication technology (ICT)-based system for older adults to use at home in order to reduce common fall risk factors such as impaired balance and muscle weakness. This article reports on usability, user-experience and user-acceptance who used the iStoppFalls system by older adults. *Methods.* In the course of a 16-week international multicenter study, 153 community-dwelling older adults aged 65+ participated in the iStoppFalls randomized controlled trial, of which half used the system in their home to exercise and assess their risk of falling. During the study, 60 participants completed questionnaires regarding the usability, user experience and user acceptance of the iStoppFalls system. Usability was measured with the System Usability Scale (SUS). For user experience the Physical Activity Enjoyment Scale (PACES) was applied. User acceptance was assessed with the Dynamic Acceptance Model for the Re-evaluation of Technologies (DART). To collect more detailed data on usability, user experience and user acceptance, additional qualitative interviews were conducted with participants. *Results.* Participants evaluated the usability of the system with an overall score of 62 (Standard Deviation, SD 15.58) out of 100, which suggests good usability. Most users enjoyed the iStoppFalls games and assessments, as shown by the overall PACES score of 31 (SD 8.03). With a score of 0.87 (SD 0.26), user acceptance results showed that participants accepted the iStoppFalls system for use in their own home. Interview data suggested that certain factors such as motivation, complexity or graphical design were different for gender and age.

5.1 Introduction

Digital gaming by and for older adults has become a popular area of research within the wider field of game studies and gerontology. There is an increasing awareness in medical and gerontology research that digital games, especially whole-body movement games, have the potential to reduce falls and increase overall health and quality of life (QoL) (Gillespie et al., 2012; Smith et al., 2011).

© Springer Fachmedien Wiesbaden GmbH, part of Springer Nature 2018
D. D. Vaziri, *Facilitating Daily Life Integration of Technologies for Active and Healthy Aging*, Informationsmanagement in Theorie und Praxis,
https://doi.org/10.1007/978-3-658-22875-0_5

Information and communication technology (ICT)-based programs offer a promising alternative method to reduce fall risk factors in older adults in their own home, using digital game-based systems and unobtrusive sensoring (Gschwind et al., 2015). The main advantage is that these serious games for health, exergames in particular, can combine exercise and entertainment for older adults (Bleakley et al., 2015).

Long-term motivation and sustainable usage of exercise and health games by older adults remains unclear (Wiemeyer, 2010). The factors which influence whether older adults accept or reject technologies in their lives are complex and diverse, include factors like gender and age (Czaja et al., 2006; Fazeli et al., 2013; Sheikh and Abbas, 2015), and differ greatly from younger technology users (Wilkowska and Ziefle, 2009). It is important to understand the technological requirements of older adults when designing high-quality systems (Keith and Whitney, 1998; Lindsay et al., 2012). Ideally older adults should be involved as "*co-designers*" in the process (Hartswood et al., 2008) to enable more detailed insights on day-to-day life aspects and interactions with digital technologies ((Lindsay et al., 2012) p.1199); and thus the best possible approach to negotiate between design and user needs.

In order to successfully design such ICT-based systems for older adults, it is crucial to understand the users' values, behaviours, attitudes, practices and technical experiences (Wulf et al., 2011) regarding digital technologies and games. Usability, user experience and user acceptance play an important role in order to collect and interpret values, behaviours and practices of users for the design of such ICT-based systems. Using a combination of qualitative and quantitative data material, this paper aims to identify factors influencing usability, user experience and user acceptance of older adults engaging with an ICT-based fall prevention system (iStoppFalls).

5.2   Methods

5.2.1   Study setup
This paper is using data from an international, multicenter study designed as a single-blinded, two-group randomized clinical trial (RCT) with a total of 153 community-dwelling older adults (n = 78 intervention group, n = 75 control group) aged 65+ from Cologne, Germany (n = 59), Valencia, Spain (n = 37) and Sydney, Australia (n = 57) (Gschwind et al., 2015); and a more qualitative-oriented Living Lab (LL) study with a total of 15 community-dwelling older adults aged 65+ from Siegen, Germany.

The iStoppFalls system used in this study consists of several technical components: 1) a set-top box with controller, 2) a mini personal computer (PC) with exergames, 3) a Microsoft-Kinect for movement detection, gesture and voice control, 4) a Senior Mobility Monitor (SMM) for mobility tracking worn around the neck, 5) a tablet PC as an alternative input and output device for the interactive television (iTV) system, and 6) an iTV. A more detailed description can be found in (Marston et al., 2015).

In the RCT, the system's effectiveness in terms of fall risk reduction and supplementary facets such as physical, cognitive and health related variables (Gschwind et al., 2014) were quantitatively analysed. The LL (Ogonowski et al., 2016) qualitatively examined the systems' suitability for integration into daily routines, of older adults', aspects of usability, user experience and user acceptance. Both, the RCT and LL, were conducted simultaneously. RCT group participants conducted a 16-week exercise program using the iStoppFalls system through the TV set in their own home (Marston et al., 2015). LL participants conducted the same exercise program and setup for 24 weeks. The extended study period of the latter was based on the LL approach which requires long-term evaluation of qualitative aspects such as the integration into daily routines (Almirall and Wareham, 2008; Schaffers et al., 2011). Nevertheless, all participants used the exact same system including exergames and assessments.

5.2.2   Study protocol
Predefined inclusion and exclusion criteria were applied to screen subjects (Gschwind et al., 2014). Ethical approval was granted by the ethics committees of the German Sport University Cologne, the Polytechnic University of Valencia and the Human Research Ethics Committee of the University of New South Wales. The most relevant requirements for participation were a broadband internet connection, a high definition TV with HDMI port and at least three meters space in front of the TV. For the LL study, an interest in preventive training (a weekly balance and strength training together with a monthly fall risk assessment) and the willingness to attend assessments or workshops at the different study centres had to be expressed. No financial compensation was offered to participants.

5.2.3   Data collection
Data from different paper-based questionnaires relating to usability, user experience and technology acceptance were collected during the respective study periods. Feasibility of the questionnaires was pretested with 10 participants and revisions were conducted according to participants' feedback. All questionnaires were distributed among the RCT and LL participants (at week 4, week 8 and week 16). In addition to quantitative questionnaires, qualitative data was collected by conducting face-to-face semi-structured interviews with a more detailed focus on usability, user experience and user acceptance. These interviews lasted between 30

to 120 minutes. For LL participants, face-to-face interviews were supplemented by observations regarding the interaction behaviour of participants with the iStoppFalls system and workshops with participants providing a platform to exchange and discuss ideas concerning the system.

5.2.4 Measures

In total, 60 participants completed paper-based questionnaires (50 RCT and 10 LL). The sample constisted of 23 male and 37 female participants with an average age of 73 years. Table 1 provides an overview of participants' characteristics.

Tab 1. Participants' characteristics

	Cologne	Valencia	Sydney	Siegen	Overall
Participants (n)	n = 15	n = 20	n = 15	n =10	n = 60
Intervention period (months)	4	4	4	6	4 to 6
Mean age (years, SD)	72.1 ± 3.6	71.5 ± 3.8	76.5 ± 4.6	70.9 ± 3.9	72.6 ± 4.0
Female (n, %)	9 (60)	13 (65)	9 (60)	6 (60)	37 (61,7)

The System Usability Scale (SUS) measures the usability of a product and consists of 10 items which are evaluated on a 5-point Likert scale ranging from 1 "strongly disagree" to 5 "strongly agree". The results are distributed on a specific scale ranging from 0 for "worst imaginable" to 100 for "best imaginable" (Brooke, 1996). SUS is an appropriate and robust usability measure with easy application for the user (Bangor et al., 2008; Borsci et al., 2009). It is frequently applied in design studies evaluating the application of interfaces (Raptis et al., 2013). Recent publications illustrate a meaningful application of the SUS in evaluation settings with older adults and within a fall prevention context (Grindrod et al., 2014; Nawaz et al., 2014).

The Physical Activity Enjoyment Scale (PACES) is frequently applied to measure the enjoyment of physical activities. Since physical activity is a core feature of the iStoppFalls system, the PACES was deemed appropriate to measure user experience. In this study, the 8-item PACES version was applied (McArthur and Raedeke, 2009; Raedeke, 2007). All items are evaluated on a 6-point Likert scale.

The responses from each participant are added up and averaged. A high score implies high enjoyment while being phyiscally active. The maximum enjoyment score for this scale is 48.

The Dynamic Acceptance Model for the Re-evaluation of Technologies (DART) is an instrument to analyze and evaluate the user acceptance of products or services without statistical analyses (Amberg et al., 2005). It provides a meta structure with four dimensions: perceived usefulness, perceived ease of use, perceived network effects and perceived costs. For each dimension individual acceptance indicators (AI) such as ease of use or attractiveness of design are defined to measure the importance and implementation perceptions of users in regard to the tested system. The definition of AI's is based on literature review and user research in the field of ICT for older adults (Chen and Chan, 2011). The users' perceived importance to an AI and the users' perceived implementation of an AI within the system are rated on a 6-point Likert scale for each indicator, with 1 "being very unimportant/totally unfulfilled" and 6 "being very important/totally fulfilled". For each AI all ratings are averaged. Figure 4 illustrates the described method via spider charts using one line for the importance and one line for the implementation. Subtracting the importance ratings from the implementation ratings provide the degree of discrepancy between the users' perception of importance and implementation within the tested system. This calculation gives a measure to explain the fulfillment of user expectations by the tested system. Values close to zero represent a high rate of acceptance from a users' perspective.

Additional semi-strucutured interviews with a focus on health, mobility, system and technology use were conducted with 40 participants. We used open ended questions in order to obtain more detailed insights on participants' perceptions. All interview participants were selected randomly, considering the predefined inclusion and exclusion criteria described above.

A combination of qualitative and quantitative methods was used (Brannen, 2005). Descriptive analyses were conducted in SPSS version 22. The SUS and PACES were applied for the whole iStoppFalls system, whilst the DART was seperately applied for those participants performing the exergames and wearing the SMM. Therefore, DART data will also be presented for the exergame and SMM. All data was structured into two different perspectives, namely age (median = 72) and gender.

Qualitative data analyses were conducted to further enhance our understanding of users' attitudes and practices. Qualitative data analysis used a content analytic approach using MaxQDA version 12 (Mayring, 2000). Coding and codes which were formed based on the interview guides were supplemented by an open inductive coding based on the overall material (Corbin and Strauss, 2014). Coding was undertaken by researchers who worked closely with the participants and were present during the intervention at the LL. They were supported by additional researchers who were not part of the LL. Empirical data was analyzed in respect of substantive data, coded interviews, transcripts, usability tests and workshops which were triangulated with additional observation notes taken right after every visit.

5.3   Results

5.3.1   Usability

Participants evaluated the iStoppFalls system's usability with an overall score of 62 (SD 15,58) which is good (See figure 2). In terms of gender, there was no noticeable difference for the perception of usability. Male participants evaluated the usability of the system with a score of 61 (SD 19,17), while female participants evaluated the usability with a score of 62 (SD 23,46). In regard to participants' age, our results illustrate that participants younger than the median age (72 years) assessed the system's usability with a score of 72 (SD 16,22) while participants older than the median age assessed the usability with a score of 53 (SD 23,64). Figure 2 provides an overview of the SUS results.

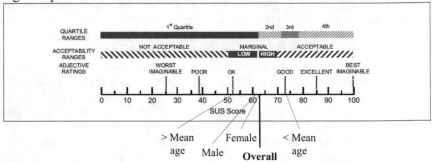

Fig 2. Overall System Usability Scale score

During the interviews, participants reported technical issues with the system: "In the beginning there were situations (…) I could have thrown the computer against the wall (…) but then, after new software has been uploaded, it worked much better, and I enjoyed playing the games again." (Siegen, Living Lab, 64 years, female).

Apart from the technical issues participants also stated that they perceived the system as user-friendly: "The system, in its current state, can almost be described as 'consumer-friendly'; especially for older adults. You always have to start from the premise that, especially older adults, struggle learning new technology – not concerning the exercises but rather pertaining to all the button pressing." (Siegen, Living Lab, 67 years, male). However, some older participants noted that the system was too complex to use without any help: "However, when I was left alone in home to use it, I realised that there were many more things than what I was able to remember. I found it a bit overwhelming, and I was afraid of not using it right. I appreciated the instructions that helped me to go through the games." (Valencia, RCT, 73 years, male).

5.3.2   User experience
The overall enjoyment of using the iStoppFalls system was evaluated with a score of 31 (SD 8,03), which means enjoyable (see figure 3). In regard to gender, male participants evaluated the enjoyment of using the system with a score of 29 (SD 7,26), while their female counterparts evaluated the enjoyment with a score of 32 (SD 8,44). Younger participants (< 72 years) assessed the enjoyment of using iStoppFalls with a score of 32 (SD 7,98). Older participants (> 72 years) assessed the enjoyment with a score of 29 (SD 7,90). Figure 3 presents the PACES results.

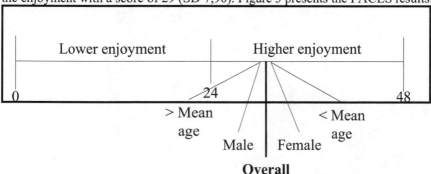

Fig 3. Overall Physical Activity Enjoyment Scale score

The qualitative content analysis showed that participants' enjoyment using the system was quite distinct. Some participants enjoyed playing the games: „*Overall I enjoyed the games, even though they were quite challenging and the Kinect didn't work properly sometimes.*" (Siegen, Living Lab, 77 years, female). Over time, other participants found the games to become boring: "*Actually, I didn't enjoy the games after some time. It just repeats over and over and gets a little boring then. If there would be an increase in the training (difficulty) then I could do it this way.*" (Siegen, Living Lab, 74 years, male).

5.3.3   User acceptance

The participants' evaluation of the acceptance of iStoppFalls is illustrated in Figure 4. The overall acceptance rate of the iStoppFalls system (SMM and exergame) was 0.87 (SD 0,26), which indicates a good user acceptance. Considering the exergame and SMM separately, the acceptance rate for the exergame was evaluated with a score of 0.96 (SD 0,28), whereas the acceptance rate for the SMM was rated with a score of 0.78 (SD 0,27). In terms of age, evaluations of participants younger than the median age resulted in an acceptance rate for the exergame of 0.44 (SD 0,18). Older participants' evaluations resulted in an acceptance rate of 0.36 (SD 0.40). For the SMM, the acceptance rate evaluated by younger participants was 0.78 (SD 0,51), while older participants evaluated the acceptance for the SMM with 0.04 (SD 0,29). In regards to gender, male participants rated the acceptance of the exergame with 0.92 (SD 0.54), while female participants' acceptance rate was 1.01 (SD 0.22). The acceptance of the SMM was 0.65 (SD 0.30) for male participants and 0.87 (SD 0.43) for female participants.

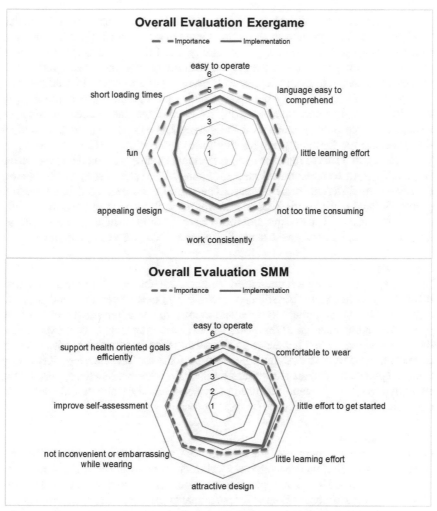

Fig 4. Overall evaluations for exergame and SMM (mean values)

In addition, the interviews revealed specific factors influencing the acceptance of the system. For instance, some participants did not like the appearance of the SMM: *"Well, it [the SMM] just does not look good. Others may think - what does she have?! And the lights as well! I used it once in my gymnastics group and immediately turned the lamp to the body to disguise it."* (Valencia, RCT, 73 years, female).

However, other participants reported liking the visualization of results: "The pendant [SMM] was something that I wore in the beginning just because you told me: I wanted to contribute to your study, but I didn't see the motivation of wearing that thing, until I saw the graphs with the results, the map, and so on. It was funny to see, and even tried to walk more outside to see if I could beat my record." (Valencia, RCT, 70 years, male). With regard to intentions to use the system after the study a participant stated: Yes, I could imagine to keep on using the system for three times a week. But the system has too be improved. Technical failures occur too frequently and loading times are too long." (Cologne, RCT, 79 years, male). In terms of the exergame participants mainly responded in a positive way, aside from some technical issues: "I really liked the variety of the games, especially the different forms of sport. The games were presented in a beautiful graphic. What I really liked on top of that was the ranking of results achieved while playing the games. I really got ambitious here. I couldn't imagine becoming so ambitious about that before the study." (Cologne, RCT, 68 years, female).

5.4   Discussion

In this paper, data from a home-based study on ICT-based fall prevention across different countries were investigated on usability, user experience and user acceptance. For the iStoppFalls system, age seemed to be an important factor for the perception and evaluation of the system's usability. In respect to user experience, the findings suggest that game design should consider gender differences in order to make the use of such systems enjoyable for male and female users. Regarding user acceptance results, the study identified tracking devices like the applied SMM as an important component, because it may considerably motivate older adults to continuously use the system and be more active.

5.4.1   Usability

Overall, participants perceived the usability of the iStoppFalls system as "good". In respect to gender, there were no noticable differences in the evaluation of the system's usability. However, regarding age, younger participants assessed the usability of the system better than older participants. As previously noted by participants during the interviews, malfunctions, long loading times or complex tasks prevented the appropriate use of the system. Referring to the results in our sample it seems that such usability issues had a stronger impact on older participants than on younger participants. These findings correspond to research investigating usability of ICT for older adults. (UNHCR, 2015) mentioned the importance of reducing malfunctions and providing easy solutions for older adults. In a different study, (Phiriyapokanon, 2011) found that complexity is a crucial factor regarding usability for older adults and should be reduced as much as possible.

5.4.2 User experience

In general, the participants enjoyed using the iStoppFalls system. The results do not show any noticable differences for the enjoyment of the system regarding gender and age. However, the interviews revealed that there seems to be a gender difference regarding the enjoyment of the system. Some male participants stated that they found the games unchallenging or that the games became boring after some time. Such statements were not made by female participants, indicating that the user experience for male and female participants differs in the case of ICT-based fall prevention systems like iStoppFalls. This coincides with the current literature on gender differences in user experience of video games (Kari et al., 2012; Sheikh and Abbas, 2015). Age differences with regard to user experience were not found in the qualitative data material. Considering gender differences for the design of exergames therefore seems to be an important factor in order to ensure good user experience for the target group.

5.4.3 User acceptance

In terms of user acceptance our quantitative results showed that the iStoppFalls system was generally accepted by the target group, as all acceptance rates were close to zero. However, Figure 4 shows some indicators for improvements of the tested prototype. The most noticable descrepancies between participants' importance and implementation evaluation were: (1) the fun factor in exergames and (2) the wearing comfort of the SMM. With regard to age and gender, there were no additional noticable differences.

The qualitative analysis was able to depict a more detailed and distinct picture of user acceptance for the exergame and SMM.

According to the interview statements participants were satisfied with the exergames. Apart from some technical issues critizised by most participants, visual aspects such as the exergame graphics seemed to be an important aspect to young older participants. (Smeddinck et al., 2013) reported similar findings in their study, suggesting that graphical design should be considered as a relevant acceptance factor for older adults. Fun in exergames might be an important factor as well, as reported by the participants during the interviews. (Brauner et al., 2013) investigated the importance of fun in exergames for older adults. In the respective study, the researchers ascertained fun as a major factor for older adults in terms of exergames. They also revealed that when exergames were fun, older adults were more likely to replay the exergames than younger people (Brauner et al., 2013).

In regard to the SMM, participants pointed out that they disliked the design of the device. They described it as appearing old and dated. Some participants even stated that they would try to avoid using it in public for that reason. Male participants in particular mentioned that they did not like wearing the SMM device like a necklace. These findings indicate that visual design and novelty of ICT technologies seem to be important acceptance factors. (Selwyn, 2004) found the desire to stay up-to-date as one of the most important factors for acceptance in his study. According to our analysis, this seems to be also relevant in the case of ICT-based fall prevention, since the SMM device (research prototype) did not seem to provide a "being up-to-date feeling" for study participants, due to its appearance. Besides feeling up-to-date the analysis indicated that gender specific aspects (wearing the SMM like a necklace) might have had an influence on the acceptance of the SMM as well. On the other hand, participants were very keen on the functionalities of the SMM device such as visualizing results based on tracked movement data. Such functionalities enabled participants to follow their own physical development and compare their results and achievements to others. These functionalities were mentioned as being very motivating by the participants. Studies investigating motivational aspects for the use of technology for physical activity by older adults illustrate that tracking and feedback of physical activity are important factors for that target group (Consolvo et al., 2006; Gerling and Masuch, 2011; Uzor et al., 2012).

5.4.4   Design implications
The analysis, presented in this article, showed that the design of ICT-based fall prevention systems for older adults should consider specific aspects with regard to usability, user experience and user acceptance. In general our quantitative and qualitative analysis revealed following design implications for exergames:
- Exergames should be easy to operate
- The used language within the exergames should be easy to comprehend
- It should require little learning effort to start using the exergames
- It should not require much time effort to use the exergames
- The exergames should work without malfunctions and have short loading times
- Visualizations and graphical design should be attractive to the target group, especially for the young older adults
- Playing the exergames should be fun for the target group
- Exergames should provide different difficulty levels

In regard to the SMM, the quantitative and qualitative analysis showed following important implications for the design of such devices:
- Activity trackers should be easy to operate
- Wearing such devices should be comfortable and convenient with regard to gender specific preferences
- The effort required to start using such devices needs to be little

- The effort to learn how to use such devices should be little
- The design of activity trackers should be attractive to the target group
- Activity trackers should provide visualizations for results to enhance user motivation

In addition to the design implications summarized above, observations and qualitative interviews revealed that graphical and fun aspects of exergames are more important for younger old adults than older old adults. With regard to gender, the design of exergames may consider different contents. Combination with wearable devices like the SMM increase the motivation of older adults to use ICT-based fall prevention systems and therefore constitute an important factor for a sustainable design of fall prevention systems.

5.5   Limitations

The questionnaires used in this study were applied in a way that the SUS and PACES covered the iStoppFalls system as a whole, while the DART differentiated between exergame and SMM allowing participants to make more distinct and accurate evaluations of system components. The SUS and PACES results therefore do not give insight into how usability and user experience were perceived for the exergame or the SMM. However, our results provided valuable qualitative information about usability and user experience aspects from participants in regard to the exergame and the SMM. Finally, the sample size of this study was too small to allow for detailed statistical analyses in terms of user acceptance. Therefore, our results can not provide significances or correlations within the collected data set.

5.6   Conclusions

The results suggest that the iStoppFalls system has good usability, user experience and user acceptance and discuss the importance of taking in to account age and gender. This paper provides important information regarding motivational and sustainable use aspects, which can be used when designing usable and enjoyable ICT-based fall prevention systems. In order to achieve sustainable system use, it will be important to consider these aspects and providing motivational factors will facilitate the acceptance of ICT-based fall prevention systems by the target group.

6 Analysis of Effects and Usage Indicators for a ICT-based Fall Prevention System in Community Dwelling Older Adults

Abstract

Falls are a serious problem in aging societies. A sedentary life style and low levels of physical activity are major factors aggravating older adults' fall risk. Information and communication technology (ICT)-based fall prevention interventions are a promising approach to counteract the fall risk of this target group. For some time now, fall prevention interventions have put emphasize to video game based solutions, as video games have become more popular and accepted among older adults. Studies show that such ICT-based fall prevention interventions significantly reduce fall risk in older adults. Nevertheless, the population of older adults is fairly heterogeneous, and factors like gender, age, fitness, sociability, and so on may influence the use of such systems. Therefore, the analysis of subgroups is a common procedure to investigate the affects of various factors on the effectiveness of ICT-based systems. Many of these studies analyze the effectiveness of the system with quantitative measures only. However, the effectiveness of ICT-based fall prevention systems always depends on the sustainable system use by the target group. Qualitative analyses is generally the prime selection to identify determining usage indicators for system usage. Therefore, it seems likely that combined quantitative and qualitative investigations will generate detailed information about system effectiveness and relevant usage indicators for respective target groups. Here, we analyze the ICT-based fall prevention system, iStoppFalls, incorporating exergames and a mobility monitor as well, targeting three aims, (1) is the system effective for different subgroups of older adults, (2) what are the factors influencing fall risk reduction in older adults using the system and are there combined effects of exergaming and activity monitoring on fall risk reduction, and (3) which usage indicators explain the usage of such a system by older adults. This paper will provide a better understanding of the effectiveness of ICT-based fall prevention for different subgroups and the indicators that determine the use of such technologies by older adults.

6.1 Introduction

In an aging population, appropriate support for prolonged independent living by community-dwelling older adults becomes increasingly important. Falls and fall-related injuries are a major factor producing impaired quality of life and increased health-care costs in older adults. Over a third of older adults aged 65 and over fall

© Springer Fachmedien Wiesbaden GmbH, part of Springer Nature 2018
D. D. Vaziri, *Facilitating Daily Life Integration of Technologies for Active and Healthy Aging*, Informationsmanagement in Theorie und Praxis,
https://doi.org/10.1007/978-3-658-22875-0_6

at least once a year. About a quarter to half of falls result in severe injuries, further falls and loss of autonomy (Kannus et al., 2005; Stevens et al., 2006). Therefore, it is important to understand falls in older adults and to develop appropriate prevention and intervention methods to reduce fall risk and falls. In relation to this, physical exercise arguably presents the most important approach for fall prevention and rehabilitation (Gillespie et al., 2012; Sherrington et al., 2008). Information and communication technology (ICT)-based approaches, it is argued, offer promising solutions to promote physical activity in older adults, by enhancing fun and motivation, long-term compliance and adherence and social inclusion (Smith et al., 2011). In particular, controller free full-body motion solutions like Microsoft (MS) Kinect offer new possibilities for the design of exergames for older adults based on full-body motion-tracking (Gerling et al., 2012; Uzor and Baillie, 2014). Newly emerging options to support fall preventive exercise by ICT are accelerometer-based mobility monitors and wearables like smart watches or other pendants which monitor older adults activities and fall risk during daily life (de Bruin et al., 2008). As with other aspects of ICT design with and for older adults, there are well known problems and issues related to the use and acceptance of such ICT-based interventions (Simek et al., 2012; Vaziri et al., 2016). Commercially available exergames and mobility monitors are seldom tailored specifically to the needs, desires, attitudes, capabilities and practices of older adults (Ijsselsteijn et al., 2007). Although some earlier research (Yardley et al., 2008) has already pinpointed this problem for non ICT-based approaches, there is still a need to identify the most important factors which contribute to adherence to exercise and efficacy of ICT-based fall prevention technology in community-dwelling older adults (Miller et al., 2014). In order to support and achieve this goal of supporting active ageing it is important to recognize the heterogeneity of older adults. Several studies have addressed the varied characteristics of this population by conducting quantitative analyses of the effectiveness of ICT-based fall prevention systems for different subgroups (Agmon et al., 2011; Bainbridge et al., 2011; Bateni, 2012; Bell et al., 2011; Bomberger, 2010; Daniel, 2012; Gschwind et al., 2015; Janssen et al., 2013; Jorgensen et al., 2013; Lamoth Claudine et al., 2011; Laver et al., 2012; Nitz et al., 2010; Pichierri et al., 2012; Pluchino et al., 2012; Singh et al., 2012; Studenski et al., 2010; Tange et al., 2012; Williams et al., 2011, 2010). Most of these studies are randomized clinical trials (RCT). However, only few comparative studies applied qualitative methods in addition to quantitative measures to investigate enjoyment and acceptance indicators of users (Bell et al., 2011; Valenzuela et al., 2016; Williams et al., 2010). The application of qualitative methods in the population of RCT's is uncommon, largely due to the assumptions behind RCT's, which have to do with scientific and statistical rigor, and hence mainly focus on the measurable effectiveness of interventions (Drabble et al., 2014; Lewin et al.,

2009; Murtagh et al., 2007; O'Cathain et al., 2013). The degree to which user experiences can be scientifically measured using such techniques is, in our view, debatable, and the significance of this lies in the degree to which the heterogeneous experiences of users may themselves impact on willingness to engage with exergames and similar technologies. Put simply, there may be both motivational (positive) and 'hygiene' (negative) factors influencing outcomes (Herzberg, 1966; Herzberg et al., 2011) and disentangling them may not be straightforward. With regard to ICT-based fall preventions, the literature provides evidence for effectiveness in reducing falls in older adults (Gschwind et al., 2014; Schoene et al., 2013; Uzor and Baillie, 2014). Main findings of the iStoppFalls RCT illustrate a significant overall fall risk reduction of 35% (intervention vs. control) in the intention-to-treat (ITT) population of the study (Gschwind et al., 2015). Nonetheless, current investigations of factors influencing fall risk reduction with iStoppFalls only deal with baseline fall risk and system adherence. Gschwind et al. (2015) for instance, conducted an initial investigation into factors related to the effectiveness of iStoppFalls. However, the literature indicates that a more nuanced understanding of risk is required. It is likely that a range of factors such as age, gender, IT-literacy, education and fall history might affect both willingness and desire to use the system and factors, which might mitigate against the effectiveness of iStoppFalls for its target population. We suggest that qualitative research provides a complementary, and valuable, additional approach to the understanding of usage. Our analysis will first evaluate the effectiveness of iStoppFalls for the subgroups gender, age, information technology (IT)-literacy, education, fallers and baseline balance. Secondly, we will investigate factors affecting fall risk reduction with iStoppFalls. In a final step, a qualitative analysis will identify indicators, which influence the ongoing, sustainable use of the system by older adults. The results will provide a more detailed insight into the feasibility, usage and effectiveness of iStoppFalls and similar technologies for community-dwelling older adults.

6.2   Related Work

6.2.1   Design of physical activity systems for older adults
The effectiveness of ICT-based tools which are intended to support preventive physical activity largely depends on technology acceptance and personal motivation, and adherence to a training or exercise plan. Accordingly, one line of discourse in recent human computer interaction (HCI) research has focused on support for motivational and acceptance factors. Within this discourse, the need for games which foster mental and physical stimulation for older adults has been recognized (Berkovsky et al., 2010, 2012; Campbell et al., 2008; Carmichael et al.,

2010; Consolvo et al., 2006; Gerling et al., 2015; Mueller et al., 2011), and considered to be "*a valuable means of re-introducing challenge in late life*" (Gerling et al., 2015). Tobiasson et al. (Tobiasson et al., 2012) describe a turn in HCI research, from a view that technology at its best is "*effortless to use*", and which "*blend[s] in with the rest of the world*", towards a discourse on how a "*more physical demanding interaction*" can be established (Tobiasson et al., 2012, p. 608). Their argument was that the inclusion of physical participation and bodily aspects in interaction design "*may yield benefits in terms of well-being, preventive health measures and perhaps also sustainability*", (Tobiasson et al., 2012, p. 614). In an ethnographic field study, Lee and colleagues have identified the opportunity to network and communicate with peer groups as key for older adults to stay involved and pursue an activity program to "*improve their health and wellness while maintaining their independence*"(Lee et al., 2012, p. 164). Carmichael et al. (2010) have identified several key aspects such as intensity, progression, feedback, and personal choice for the sustainable and lasting participation of older adults in an ICT-related physical activity program (Carmichael et al., 2010, p. 287). Ballegaard et al. (2008) emphasized the importance of designing ICT and healthcare technology in a way that it "*fits the routines of daily life and thus allows the citizens to continue with the activities they like and have grown used to – also with an aging body or when managing a chronic condition*". This research draws on basic findings from studies in psychology, gerontology and health care. Martin et al. (2000) provide an overview on key factors related to exercise adherence, summarizing "*(a) individual factors (including psychological variables and behavioral skills), (b) treatment factors, (c) interpersonal factors, and (d) environmental factors*" as being central for the prediction of adherence to trial requirements (Martin et al., 2000, p. 198). Chao and colleagues have similarly identified "*goal setting, self-monitoring, implementing decision-making models, modifying cognitive thoughts during activity, and increasing social support*" as useful strategies for the promotion of adherence to physical activity (Chao et al., 2000, p. 214). In line with these findings, Geraedts et al. (2014) reported that "*the personal tailoring of the exercise program as well as the individual feedback based on actual objective performance*" is central to a good adherence and effectiveness of an individually tailored home-based exercise program for frail older adults.

6.2.2   *Exergames and older adults, ICT and fall prevention*

In recent years considerable research has been undertaken in the area of exergames to combine serious gaming and health-protective exercising (Corti, 2006). Serious games have the purpose of influencing end-users to *play* for health reasons (Göbel et al., 2010). Nevertheless, fall preventive exergames for older adults are underrepresented in the existing literature. Early attempts have, even so, shown positive results regarding the acceptance of interactive videogames for rehabilitation of older adults (Betker et al., 2006; Studenski et al., 2010; Vaziri et al., 2016). Fall preventive games and exercises, it is argued, should cover a wide range of physical tasks (like balance and strength) delivered in numerous formats, some of which are likely to result in greater reductions in falls than others (Gillespie et al., 2012; Lai et al., 2013; Schoene et al., 2013; Sherrington et al., 2008). Building on these insights, ICT-based exergames can be seen as a promising approach to promote older adults' health and fall-preventive activities (Schoene et al., 2011; Smith et al., 2011). Additional advantages of whole-body movement Kinect technologies are integrated biomechanical modeling and advanced reasoning (Gerling et al., 2012) which can be used for fall prediction (Tiedemann et al., 2008). Further, it offers users and researchers the ability to track fall risk over time. Different solutions are offered, based on principles of scientific evidence, fun and motivation, long-term compliance/adherence and social inclusion (Smith et al., 2011; Yardley et al., 2008) for older adults living independently in their homes. A long-term study of exergames aiming to overcome low adherence to rehabilitation exercises has shown the positive potential of exergames in older adults' home settings (Uzor and Baillie, 2014). The results indicated that adherence to exergames was superior compared to regular exercise based on explanations written in a booklet. In general, the adherence to regular home-based exercise programs that do not provide personal supervision and support is often poor (Day et al., 2002; Freiberger et al., 2007; Lord et al., 2005; Robertson et al., 2001; Steinberg et al., 2000; Stevens et al., 2001). Even in case of personal supervision, adherence rates are not always satisfying, mainly for reasons of efforts and costs to attend groups (Schutzer, 2004; Whitehead et al., 2006; Yardley et al., 2006). Therefore, exergames are a promising approach to increase adherence in fall prevention programs. Apart from adherence, exergames can address additional aspects in the social lives of older adults. Gerling et al. (2015) conducted a three-month study with videogames in two long-term care facilities, where results indicated positive influence in late life of older adults by means of *"re-introducing challenge in the lives of older adults"* and satisfying *"their need for autonomy"*.

6.2.3 Activity monitoring in older adults (wearables)

Functional status is one of the most important factors affecting quality of life and health care utilization in old age (Ferrucci et al., 2000). Therefore, it is of importance to generate valid measures of physical activity and physical functioning in older adults to support strategies which aim at maintaining functional status (de Bruin et al., 2008). The use of wearable sensors that monitor the activity of older adults can be helpful to support interventions which increase daily physical activity and adherence to home-based physical activity and/or exercise programs (de

Bruin et al., 2008). For instance, the general recommendation of walked distance for older adults is 6000 steps per day, which converts to roughly 4.5 km (Tudor-Locke, 2002; Tudor-Locke and Bassett, 2004). Devices which track data regarding walked distance, gait, balance and strength can be worn around the participants' neck or wrists. The resulting data can be processed at regular intervals, and identified trends concerning mobility, physical activity, exercise and fall risk can be provided to the participants to enhance motivation (Brodie et al., 2015; Regter-schot et al., 2014, 2015).

6.2.4   Target groups for ICT-based fall prevention

Subgroup analysis is a common procedure identified in current literature to measure the effectiveness of fall prevention in RCTs in what is an otherwise fairly heterogeneous population. Factors such as gender, age, fitness, sociability, and so on, may all affect participation rates for a variety of reasons. The reasons for such variations may not always be immediately apparent and repay analysis. For this reason, data is collected about the target population as a whole which, when grouped on the basis of the factors identified allows for comparison. Such analysis has typically been applied for traditional and ICT-based fall prevention interventions (El-Khoury et al., 2015; Gill et al., 2016; Gschwind et al., 2015; Neyens et al., 2008; Rapp et al., 2008; Schoene et al., 2015). Our aim here is to apply the same analytic procedures (subgroup analyses) to understand better motivations for and barriers to, longer term participation. Most important reasons for the participation of older adults in fall prevention programs which include strength and mobility components are improvement of health, fun and independence (Yardley et al., 2008). Another important motivator for exercise participation is social interaction (Cantwell et al., 2012). However, participation and sustainability of older adults in prevention programs remains low and is still a challenge (Brox et al., 2011; Phillips et al., 2004). When and if technology enters the scene, motivation might become even lower as additional HCI problems may arise based on the different technological prerequisites of older adults. Czaja et al. (2006) identified different factors influencing technology use by older adults like education, socio-economic status (SES), perceived use as well as daily life attitudes and practices. Jakobs et al. (2008) found gender, age, health status and social contacts as important factors influencing the use of ICT by older adults. As we have suggested, these various factors can be understood separately as being to do with both motivation (positive features) and 'hygiene (negative features). Subgroup analyses, which separate the sample population into groups, which may have differing motivations and may experience different barriers, are therefore pertinent. Researchers commonly apply these factors to investigate the potentially variable effects of fall prevention programs on various subgroups of older adults.

6.3   Methodology

For the analysis, we chose a mixed methods approach combining quantitative and qualitative methodologies (Brannen, 2005; Mertens and Hesse-Biber, 2012). The evaluation of the effectiveness of the prototype system for reducing falls in the considered subgroups required us to analyze quantitative data regarding participants' fall risk development throughout the study. However, in order to reveal and understand better the indicators determining the quality of their experience and its relationship to use of the system, qualitative analyses seemed an appropriate additional method.

6.3.1   Study Design

In total, 153 community-dwelling older adults participated in this international, multisited two-group randomized study (78 intervention group; 75 control group). The study included three different sites: Germany (Cologne), Spain (Valencia) and Australia (Sydney). The intervention group performed a 16-week unsupervised exercise program at home. The scheduled exercise dose was 180 min per week during at least three sessions (Sherrington et al., 2008, 2011). In addition, the system offered an interactive fall risk assessment and the Philips senior mobility monitor (SMM), a research prototype that monitors activity and mobility during the conduct of daily life. The system was provided to the intervention group participants without any charges. Control group participants were not given any technological artefacts, but instead were instructed to follow their usual habitual exercise schedules and received a booklet on general health and fall prevention. For more detailed information on the applied system and methodologies see (Gschwind et al., 2014).

Ethics committees in the study centers involved gave their ethical approval for the RCT. Every participant signed a written informed consent prior to inclusion and was required to meet specific eligibility criteria: (1) aged 65 years and older, (2) living in the community, (3) able to walk 20 meters without a walking aid, (4) capable of watching TV with or without glasses from 3 m distance, and (5) enough room at home for system use (3.5 m²). Participants were excluded from the study if (1) language skills for understanding study procedures were insufficient, (2) cognition was impaired, and (3) medical conditions precluded participation in a regular exercise program. Cognitive impairment was tested with the Mini-Cog™ test (Borson et al., 2000). In addition, participants were required to get medical approval and use a chair when exercising for safety reasons. In our sample, significant ethnic differences did not exist. A more detailed description of participants, including ethics, randomization and blinding can be found in (Gschwind et al., 2014). Figure 5 depicts the participant flow chart of this study.

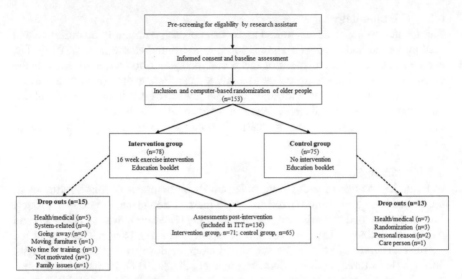

Fig 5. Participant flow chart

During the research project, a MS Kinect-based interactive television (iTV) system for older adults was developed and evaluated (Gschwind et al., 2014). It includes three different exergames that mainly support balance training (walking, stepping and leaning), strength training with various difficulty levels, and a fall risk assessment with discrete measuring technologies and adaptive assistance functions (see Figure 6). The system was connected to the TV in the participants' living room. Different input devices like a tablet-PC, speech recognition or gesture control were available to simplify usage.

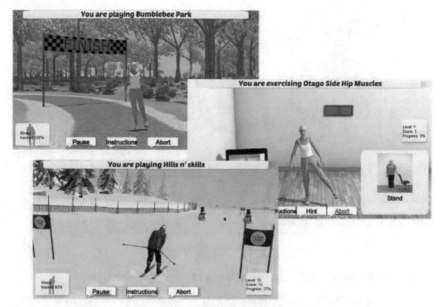

Fig 6. Balance games (left) and strength training (right)

Based on recommendations for exercise for preventing falls in older adults by Sherrington et al. (2011, 2008), intervention group participants performed a 16-week exercise program. Participants were recommended to conduct at least three balance sessions of about 40 minutes each and at least three muscle strength sessions of about 20 minutes each per week. The system provided three different balance games for walking, weight shifting, knee bending, and/or stepping in different directions. Once a participant reached a higher exergame level, the system added cognitive tasks targeting executive function and working memory. Strength exercises for the lower extremities were based on the strength exercise component of the Otago exercise program (Campbell et al., 1997). Participants were recommended to conduct 2 to 3 sets of 10 to 15 repetitions and to rest for 2 minutes in-between (Gschwind et al., 2014).

The system provided visualizations for achievements and results collected during exergames to the users. Figure 7 illustrates (from left to right) a scoring model for fall risk assessment, exergame performance evaluation and training effects visualization over time.

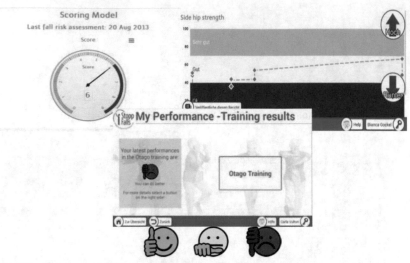

Fig 7. Exergame and fall risk assessment results visualization

During the 16-week exercise program, a pendant sensor was worn on participants' chests for data collection. Three-dimensional acceleration and air pressure data were continuously recorded on a micro secure digital (SD) card equipped inside the sensor. At regular intervals, the collected data was sent to the system for processing. Information was derived regarding wearing time, distance walked, amount of chair rise transfers and chair rise transfer peak power, previously associated to strength, balance, functional status and falls risk (Brodie et al., 2015; Regterschot et al., 2014, 2015). In contrast to the more specific exergame training, participants were not instructed to walk more while using the SMM. iStoppFalls visualized the tracked SMM activity data of users in diagrams, for instance the covered distance while wearing the SMM (Figure 8).

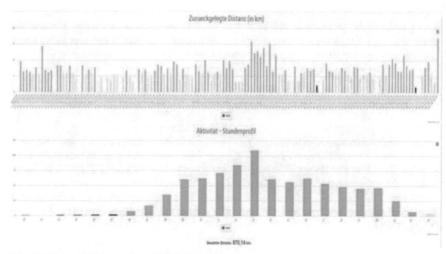

Fig 8. Visualization of covered distances

Additionally, iStoppFalls provided visualizations of covered distances, transferred into a respective radius on a google map, presenting the respective hometown of users (see figure 9).

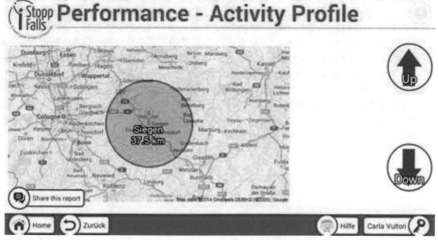

Fig 9. City comparison of group activity performances

The Google maps radius summed up the activity data from all users participating in one of the study centers (Cologne, Valencia, Sydney), visualized them in the system and thereby encouraged competition between study sites. This data was available to all participants across the study centers. In addition, iStoppFalls provided an infrastructure that allowed participants to share their achievements and high scores with the community (See figure 10).

Fig 10. High scores (on the left) and achievements (on the right) in iStoppFalls

6.3.2   Data collection
Participants were assessed at baseline (0 weeks), after 8 weeks and at the end of the intervention period (16 weeks). Accounting for dropouts, 136 participants were assessed after 16 weeks. As part of our study protocol, we collected information on fall risk, physical, cognitive, socio-demographic characteristics and medical history by applying different kinds of assessments and self-report questionnaires. These data were subjected to quantitative analysis (Gschwind et al., 2014).

For the estimation of the individual fall risk, the Physiological Profile Assessment (PPA) was applied. The PPA is based on five sensorimotor tests including contrast sensitivity (Melbourne Edge Test (MET)), peripheral sensation, balance, hand reaction time (HRT) and lower extremity muscle strength (Lord et al., 2003). Figure 11 illustrates how the PPA results can be interpreted. More detailed information on the physical outcome measure selected for this subgroup analysis is provided by (Gschwind et al., 2014).

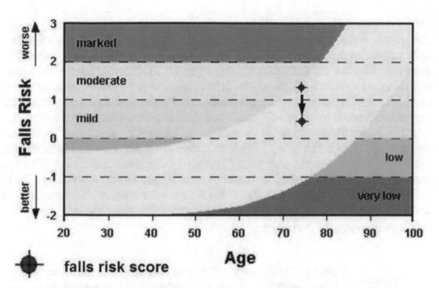

Fig 11. Virtual fall risk reduction of a 74-year-old user from 1.3 to 0.3 as measured and calculated by the PPA

In respect of IT-literacy, participants were asked for the frequency of computer use. Possible answers ranged from 1 "more than once a day" to 6 "I normally do not use a computer". The frequency of falls and adverse events were monitored with monthly diaries. In the case where diaries were not returned, participants were contacted (Gschwind et al., 2014). System usage data (e.g. adherence to exergames) was continuously collected for the exergame and the SMM device.

In addition to the study protocol, several face-to-face interviews, workshops, observations and focus group discussions were conducted. In terms of observations, a research assistant visited participants at home and observed their interaction with the system. Regarding the interviews, 12 semi-structured interviews of 30-60 minutes each were performed with intervention group participants. Interviews were carried out at participants' homes across all study centers. Furthermore, workshops with a focus on usability, accessibility and user experience were executed at monthly intervals with randomly selected intervention group participants.

6.3.3   Data analysis

Subgroups: A review of relevant scientific literature (Czaja et al., 2006; Jakobs et al., 2008; Phillips et al., 2004; Yardley et al., 2008) gave us initial insight into the potential effectiveness of ICT-based fall prevention systems, such as the iStoppFalls system, in reducing falls for specific subgroups. The quantitative data collected during our study, gave us additional insight, which we now describe. The analysis will focus on the subgroups age, gender, IT-literacy, education, fallers and balance (median cutpoints). The subgroups *age* and *gender* describe socio-demographic characteristics of participants. *IT-literacy* and *education* define the experience level in terms of ICT usage and the educational background of participants in years. Here, education is used as indicator for participants' socioeconomic status (SES). The group of *fallers* includes participants who did or did not experience falls in the past year prior to the study. *Baseline balance* describes the participants' individual anterior, posterior, medial and lateral sway (Lord et al., 2003) at the beginning of the study. Table 2 shows the baseline characteristics of the subgroups. There were no significant differences between groups.

Tab 2. Subgroup characteristics

Subgroups	Intervention Group n=78	Control Group n=75	Baseline comparison (p)
Age mean (SD)	74.71 (±6.66)	74.65 (±6.03)	.856
Gender female, n (%)	43 (55.8%)	50 (66.7%)	.144
IT-literacy mean (SD)	2.1 (±1.41)	2.4 (±1.54)	.174
Education in years mean (SD)	11.48 (±4.78)	11.07 (±4.19)	.723
Balance (SD)	841.37 (±78)	796.67 (±70)	.672
Fallers, n (%)	25 (32.1%)	26 (34.7%)	.684

System usage factors: One important pillar for the effective ICT-based reduction of fall risk in older adults is sustained usage of the system. One part of our analysis therefore focused on system use, based on quantitative data, by all participants in order to explain the effects in fall risk reduction. System use was measured by exergame adherence and walked distance. *Walked distance* indicates the activity level of the intervention group throughout the study. This includes activity, such as walking, while using the SMM as well as being active outdoors. *Exergame adherence* stands for the time per week participants played the iStoppFalls exergames during the study.

In addition, we investigated whether there was a combined effect of using the exergame and SMM device on fall risk reduction. Therefore, we pooled exergame adherence and walked distance into four different post-hoc groups: 1) Control group (no exergame AND no SMM); 2) Low Adherers (exergame adherence < 90 min AND walked distance < median); 3) Medium Adherers (exergame adherence > 90 min OR walked distance > median); 4) High Adherers (exergame adherence > 90 min AND walked distance > median). In accordance with the literature, we split the exergame adherence into one group below an adherence of 90 minutes and one group above an adherence of 90 minutes (Simek et al., 2012)

The evaluation and comparison of subgroups was not planned during the study design phase (Gschwind et al., 2014) and therefore is exploratory. Effectively, we did a post-hoc analysis focusing on correlation between each subgroup and usage factor.

Statistical analysis: Repeated ANOVAs and Student t-tests were applied to determine the intervention effects on the subgroup's fall risk (PPA), compared to the control group, and to explain the effects of system usage on fall risk reduction. In order to compare equal sized subgroups, we split most subgroups by their median value. As *gender* and *fallers* are already dichotomous, a median split was not applicable for these subgroups. All analyses were performed with SPSS version 22 (SPSS, Inc., Chicago IL). Dropouts were not included in this study (per-protocol analysis).

Qualitative analysis: The results from the quantitative analysis provide a detailed picture on the effectiveness of iStoppFalls for subgroups and how system use affects fall risk reduction in older adults. However, the quantitative analysis does not fully reveal the factors and indicators, which determine the underlying reasons for differing levels of system use by the target group. Such information, we suggest, requires thorough qualitative analysis. Therefore, we examined the qualitative data material (audio notes, text notes, videos), collected in observations and workshops with a focus on usage indicators for using iStoppFalls. In addition, we analyzed 12 semi-structured interviews with open questions focusing on e.g. perceived benefits, problems and usage of iStoppFalls. All interviews were conducted and, completely transcribed by, trained research assistants. Based on the transcripts, three coders performed a deductive qualitative content analysis (Mayring, 2003) of the data material and generated principal categories. Coders structured the data material independently of each other in the first instance. Coding variances were discussed and eliminated by adding, editing or deleting codes, based on the discussion outcomes. The final code system covered categories relating to use of media and technology in general, motivation to use media and technology for fall prevention, interaction with the system, general and technical problems during system use, recommendations for improvements and individual evaluation of the system. Based on the coded data material we derived iStoppFalls usage indicators relevant for the target group. For the analysis, coders used the software application MAXQDA. All names provided are pseudonyms.

6.4 Results

This section illustrates the results of the subgroup analysis with regard to system effectiveness. Tables 3 and 4 describe the results of the quantitative analyses.

6.4.1 Effectiveness of the system for subgroups

Age and gender: In relation to age, the iStoppFalls system did not have any significant three-way interactions on fall risk. Post hoc two-way analyses, however, revealed a borderline significant effect for younger participants for fall risk, compared with the control group (p=.068). In terms of gender, there was a significant three-way interaction between male, female and control group participants (p=.046). Post-hoc two-way analyses showed a significant larger effect for female participants, compared to the control group (p=.017).

Education and IT-literacy: A significant three-way interaction between low IT-literacy group, high IT-literacy group and the control group could be shown (p=.019). Two-way analyses illustrate a significantly larger decrease in fall risk for the high IT-literacy group, compared to the control group (p=.006). In terms of education, our results do not show any significant three-way interactions for fall risk. However, two-way post-hoc analyses, comparing the high education group with the control group, revealed a borderline significant larger effect in favor of the high education group for fall risk (p=.059).

Baseline balance and fallers: Our analyses showed a significant three-way interaction between low balance group, high balance group and control group (p=.020). Two-way post-hoc analyses showed that the high balance group had a significant larger decrease in fall risk in comparison to the control group (p=.008). Compared to low balance group, the high balance group shows a borderline significance in reducing their fall risk (p=.053). In regard to participants who experienced falls prior to the study, results show a borderline significant three-way interaction between previous fallers, non-fallers and control group (p=.072). Comparing the non-fallers group with the control group, two-way post-hoc analyses illustrate a significant larger effect in favor of the non-fallers group for fall risk (p=.041).

Tab 3. Fall risk development for specific subgroups

Subgroups	Median	Intervention Group				Control group		Post hoc comparisons			
		<median (male, faller)		>median (female, no-faller)				p-value (three-way)	p value (two-way)[1]	p value (two-way)[2]	p value (two-way)[3]
		Pre	Post	Pre	Post	Pre	Post				
Age in years	74	.47	.14	1.01	.71	.42	.39	.108	*.068**	.128	.887
Gender	N/A	.43	.25	.83	.41	.43	.40	**.046**	.621	**.017**	.190
IT-literacy score	2	.53	.31	.80	.17	.42	.39	**.019**	.227	**.006**	*.064**
Education in years	11	.69	.38	.63	.30	.43	.40	.100	.268	*.059**	.879
Baseline fall risk score	.540	-.11	-.14	1.34	.78	.42	.39	**.001**	.976	**.001**	**.003**
Fallers	N/A	.94	.62	.53	.20	.43	.41	*.072**	.110	**.041**	.979

Note: [1] comparing below median group (male, fallers) and control group / [2] comparing above median group (female, no-fallers) and control group / [3] comparing above median group and below median group

6.4.2   Effect of system usage on fall risk

Exergame adherence: Concerning system usage factors our results revealed a significant three-way interaction between low adherence (<90 min, n = 53), high adherence (>90 min, n = 18) and control group (p=.044). Two-way post-hoc analyses showed that the high adherence group had a significant larger fall risk reduction, compared to the control group (p=.031). A weak borderline significant effect on fall risk reduction could be identified for the low adherence group, compared to the control group (p=.099).

Walked distance (SMM): The median walked distance was 12.96 km per week and participant in the intervention group. About 20 percent of the intervention group participants exceeded published recommendation of 6000 steps a day (Tudor-Locke, 2002; Tudor-Locke and Bassett, 2004). There was a significant three-way interaction between the low walked distance group, high walked distance group and control group (p=.023). Two-way post-hoc analyses illustrated that the high walked distance group had a significant larger decrease in fall risk compared to the control group (p=.008), and a borderline significant larger decrease in fall risk compared to the low walked distance group (p=.066). Table 4 illustrates the changes of PPA mean values for the system usage factors as described above. Baseline comparison for the intervention groups <median and >median shows no significant differences.

Tab 4. Mean values for PPA fall risk pre and post intervention as related to the system usage factors

System usage factors	Intervention group				Control group		p-value (three-way)	Post-hoc comparisons		
	<90 min, <median		>90 min, >median					p value (two-way)[1]	p value (two-way)[2]	p value (two-way)[3]
	Pre	Post	Pre	Post	Pre	Post				
Walked distance	.81	.65	.56	.06	.42	.39	**.023**	.462	**.008**	*.066**
Exergame adherence	.55	.34	.84	.39	.42	.39	**.044**	*.099**	**.031**	.224

Note: [1] comparing below median group (male, fallers) and control group / [2] comparing above median group (female, no-fallers) and control group / [3] comparing above median group and below median group / * borderline significance

Combination of Exergame and SMM: There was a significant continuous overall effect for the reduction of fall risk between all groups (p<.045). Participants who adhered more (>90 min) to the exergame and walked more distance with the SMM, benefitted most from the intervention in terms of fall risk reduction (line to the right in Figure 12). This effect persisted after normalizing for baseline fall risk (p<.039). Figure 12 shows the 95% confidence intervals (CI) for the respective fall risk reduction of the predefined groups.

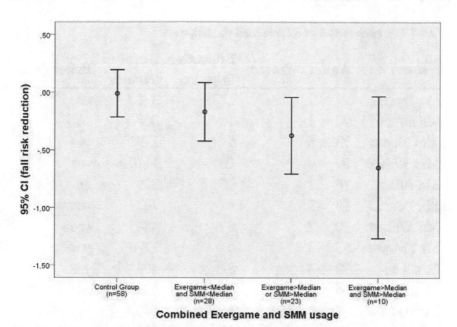

Fig 12. 95% confidence intervals of the fall risk reduction (PPA) for post-hoc defined groups (combination of exergame and SMM)

6.4.3   Usage indicators for iStoppFalls

According to the qualitative content analysis, nine different topics of indicators determining the use of iStoppFalls by older adults were revealed. The results will be described in this section. Table 5 provides an overview of interviewed participants and their characteristics.

Tab 5. Characteristics of interviewed participants

Name	Age	IT-literacy	Education in years	Baseline balance	Faller (y/n)
Mr. Davis	72	1	12	378	no
Mr. Bryan	70	1	18	368	no
Mrs. Bryan	73	3	5	2793	yes
Mrs. Wilson	69	4	12	1170	yes
Mr. Adam	78	1	16	399	no
Mr. Peters	84	3	4	792	yes
Mr. Walters	70	2	8	484	yes
Mrs. Meyers	73	3	4	1170	yes
Mrs. Lee	72	4	8	408	yes
Mrs. Bernard	66	1	16	693	no
Mr. Allen	85	3	16	640	yes
Mrs. Charles	70	1	12	323	no

Level of difficulty (individual progression): The system, participants suggested, should support the individual progression of the participant who uses the system. While it automatically works in the background, the participant should always be in control if he wants to increase (or decrease) the level of difficulty. If the level of difficulty increases too fast, participants are nervous about keeping on playing: *"For instance, the bar game is now almost impossible to play! I am afraid of not being able to play the rest of the games if this keeps on that way."* (Mr. Davis). A female participant experienced the same challenges while playing the different games: *"...although I still think that they are sometimes too difficult to me, if I don't have the support of another person."* (Mrs. Wilson). Another important factor are game goals (such as high scores), and these game elements should be appropriate for older adults: *"I'm an achievement person. I want to count points."* (Mrs. Charles).

Easy to interact (operating options): The current prototype provided different opportunities to use the system (gesture and voice control, remote control and an app for the tablet), but participants required time to figure out which operating option is adequate for them: *"But I would need more time to learn use it, and perhaps the system might be made simpler."* (Mrs. Bryan). Also one participant stated that *"Games for older people like me, who are not very accustomed to use computers, would be more motivating if they were easier to use.*" (Mrs. Wilson). Even the SMM, which was controlled by pressing only one big button, still requires some improvement. For example: "*It [SMM] should be miniaturised*". (Mr. Scott).

Error tolerance: During the first few weeks, participants had to cope with many errors and issues regarding different aspects of the system (e.g. problems with the recognition of the camera, connectivity problems with the exercise server). The system should, they felt, be improved to enable an unproblematic progress, even when issues occur. These problems caused participants to stop using the system: "*Over the program I still enjoyed the [balance] games when the system worked, but I stopped the strength exercises as it [the camera] wouldn't sense what you were doing*". (Mr. Adam).

New activities are important to keep the motivation high: It is important to challenge the participants with new activities such as different games or more demanding interactions. Participants requested these extensions of the current system: "*Perhaps combining new easy games with the difficulty of the old ones.*" (Mr. Davis). Another request was to expand the exercises using the whole body: "*Maybe a little bit more challenging. Not only raising the legs, but rather something like yoga or some gymnastic exercises. Just walking through the park is too simple.*" (Mr. Davis). For some participants the games become boring after some time, due to the lack of new content and activities: "*I found some of the exercises rather boring*" (Mrs. Bernard).

Playing together: Another usage indicator refers to the possibility for participants to train together. They might, for instance, use the same system in front of the TV or over the internet in their respective households: "*It would be more fun if we could play together with two characters at the same time.*" (Mr. Bryan). A married couple played one after another and stated, "*In the end I was even teased by the game and competing with my husband.*" (Mrs. Wilson).

Feedback mechanisms: The results of the training should, it was felt, be presented in a descriptive and positive way. Participants should have the opportunity to see what they achieved in the last sessions and even go further with more detailed information: "*I would have liked more timely and complete information about the exercises I do.*" (Mr. Bryan). Controlling their activities over a period of time was a strong motivational factor as stated by one participant: "*I like to see myself improve. If I check the results and realize that I did not perform too well the other day, I get motivated to improve myself. That is what the system is for, isn't it?*" The results right after the exercises were seen as positive feedback: "*This is really interesting, how many calories were burnt and how many kilometers were made. The SMM is not bad, as a kilometer counting device.*" (Mrs. Bryan). The feedback also generated perceived benefits for the users, as one participant stated:

"Using the system definitely helped me. It showed me where my weak spots are and where I need to improve." (Mrs. Charles). In the same context, another participant said, *"Using the system was beneficial for me. I notice that my balance has improved a lot. I would like to continue using the system."*(Mr. Allen).

Integration of the system in the daily routine: Most of the participants valued the advantage of the system for home-based training: *"I like intimacy of doing exercises at home, and the opportunity of learning exercises, routines and receiving advice to be healthy and avoid falls. I think it makes me feel safer than before."* (Mrs. Bryan). They could, they thought, integrate the training system in their daily routine and exer¬cise whenever it fits with their daily activities: *"After I finished my morning coffee, I thought that I can start with the training. I integrated the session in my daily routine and sometimes I also trained in the afternoon again."* (Mrs. Lee). Another participant experienced the same: *"Yes, I noticed a benefit of using the system. These games, which make you stand on your toes or let you walk through the park, they really force you to lift your legs properly. I have adopted this to my daily routine."* (Mr. Walters).

Creating additional value by using the system: Besides the physical training, the system might also create additional value. Participants liked the fact that they could learn new information about health, falls and how to improve their fall risk: *"I have also learnt many things about falls and health."* (Mr. Bryan). Also, the same experience was reported by another participant: *"I want to be healthy, and any support for that is welcome. I am curious about videogames."* (Mr. Davis). Other participants valued using the system for improving their awareness of physical condition: *"Yes! You definitely feel the physical advantages, when you apply the system"* (Mr. Peters). In the same context, participants said that, *"Since then [using the system] I am much more aware of the way I walk, and try to walk properly most of the time."* (Mrs. Meyers) or *"Every time I use the system I feel very good afterwards."* (Mr. Adam). For other participants the additional value was to take part in the iStoppFalls project and meet new people: *"I just want to take part in the project. See how it goes and how my performance is in relation. [...] the new contacts, to be part in something. This is my incentive."* (Mr. Allen).

Combination of different devices: The iStoppFalls project provided two devices (Kinect-based exergame and SMM) for different usage contexts (indoor training and outdoor activity tracking). A combination of these devices may have encouraged the participants to train at home and be more active outdoors. Some even preferred the activity-tracking device over the indoor training: *"I prefer the SMM over the Exergames - put it on your neck and that's it!"* (Mr. Davis) The system helped them to be aware of their own physical condition, undertake responsibility and train whenever they want to: *"I liked that I could use iStoppFalls in my apart¬ment, without going to the gym. I can decide on my own, how my physical condition is. It is my duty to make some¬thing out of it."* (Mr. Peters). The feedback from the devices was also a motivational factor for participants: *"I was motivated to check, if the results are available. Seeing some kind of progress helped me to keep on with the trai¬ning."* (Mrs. Wilson). The combination of several devices had an impact on the motivation of the participants: *"I really like to compete and my main interest is in my city's position in the performance chart. I would go for another walk and put my device [SMM] on."* (Mr. Adam).

6.5 Discussion

Targeted exercises for strength and balance are able to reduce falls and fall risk in community-dwelling older adults (Gillespie et al., 2012; Sherrington et al., 2011). However, a positive effect can only be achieved by means of a sufficient exercise dosage and sustainable training with a good adherence over a longer period of time (Phillips et al., 2004), and thus the motivation and practices (Wulf et al., 2015c) of the participants play an important role. A promising solution to support older adults in this regard are ICT-based fall prevention systems like iStoppFalls or others (Brox et al., 2011; Schoene et al., 2013; Uzor and Baillie, 2014). However, there still exists a need for researching concepts and solutions to enhance sustainable use of such programs by older adults (Broekhuizen et al., 2012; Phillips et al., 2004; Yardley et al., 2008). Designing appropriate fall preventive interventions based on exergames or wearables for older adults still remains a challenge despite a number of available concepts like persuasive design (Fogg, 2002, 2009), including gamification and serious gaming, as well as concepts like embodiment (Merleau-Ponty, 2002; Phillips et al., 2004) and self-efficacy (Bandura, 1977, 1993). In our study and in the underlying software design, we incorporated all these concepts, and accordingly, results in terms of system effectiveness and usage indicators will be discussed in the following sections.

6.5.1 Discussion of subgroup results

Age and gender: For the subgroup age our results showed that there was no significant effect on the reduction of fall risk between younger and older participants. A possible conclusion is that the fall risk reduction of the ICT program may be relevant for all age groups within the older population (65 years and over). In contrast, our quantitative analysis yielded a gender effect on fall risk reduction. According to these results, female participants seem to be able to reduce their fall risk almost twice as much as their male counterparts over the duration of the study period. An extended multifactorial analysis, including baseline fall risk as a covariate, illustrated that only the baseline fall risk had a significant effect on fall risk reduction, while the gender effect disappeared. Finally, it seems that the overall fall risk reduction induced by the ICT-based fall prevention system is equally effective for male and female older adults. The conclusions made for age and gender subgroups are confirmed by many other studies investigating effects of ICT-based fall prevention on older adults (Bainbridge et al., 2011; Bateni, 2012; Daniel, 2012; Pichierri et al., 2012; Pluchino et al., 2012; Singh et al., 2012; Williams et al., 2010).

IT-literacy and education: The results for this section indicated that participants' education had no significant effect on the reduction of fall risk, and thus the program may be effective independent of individual education levels. In contrast, several studies investigating the impact of SES on falls in older adults found a significant influence of education on the reduction of falls by means of non ICT-based fall prevention programs (Fabre et al., 2010; Gill et al., 2005; Matsuda, 2010; L. Yardley et al., 2008). In our quantitative analysis, we did not find similar results, suggesting that education may not be directly associated with the effectiveness of ICT-based fall prevention interventions. It should be noted that the majority of our participants had moderate to high levels of education, indicating that all were from a relatively affluent socio-economic background. In terms of a more specific IT-literacy our results imply an overall effect on the achieved fall risk reduction in favor of higher IT-literacy participants. The effect in fall risk reduction for participants with higher IT-literacy was eliminated after including baseline fall risk as a covariate, suggesting that IT-literacy seems not to influence fall risk reduction directly. Nonetheless, increased IT-literacy eases the access to and use of ICT-based technologies (Charness and Boot, 2009; Selwyn et al., 2003) making it an important and probably mediating factor for the effectiveness of ICT-based fall prevention systems.

Baseline balance and fallers: In terms of baseline balance, quantitative data indicated that participants with a worse prior balance benefited most from using the iStoppFalls system in terms of fall risk reduction. Participants with better balance only slightly improved their fall risk. This implies that iStoppFalls seems to be effective for the main target group of the system, i.e. older adults with a worse balance and therefore at higher risk of falling. These findings are consistent with current literature and research, as many studies indicate that high risk populations benefit more from fall prevention interventions, while low risk populations benefit less (Gillespie et al., 2012; Health Quality Ontario, 2008; Karlsson et al., 2013; Tinetti and Kumar, 2010). Quantitative analysis did not produce evidence that fall history directly affects fall risk reduction with iStoppFalls. Therefore, we can assume that this target group equally benefits from the iStoppFalls system.

6.5.2   Discussion of system usage factors

The quantitative analysis of usage factors demonstrates how far effectiveness (fall risk reduction) evolved through continuous use of the system. In addition, the analysis revealed that the combination of indoor home training with exergames, together with outdoor activities tracked by the SMM seems to amplify the effectiveness of fall risk reduction in older adults.

Exergame adherence: PPA values demonstrated that a higher exergame adherence improved fall risk more effectively than lower adherence and control. However, the literature suggests that an exergame adherence of 180 minutes per week over a course of four months (50 hours in total) might be necessary in order to achieve significant improvements in fall risk (Sherrington et al., 2011, 2008). Our study provided data for participants who exercised more or less 90 minutes per week, but even so significant effects on fall risk reduction were seen. This implies that, especially if long-term adherence could be achieved beyond a 4 to 6 month period, an exercise dose of 60 to 90 minutes might be sufficient to have an optimal effectiveness of the iStoppFalls system towards reducing fall risk.

Walked distance (SMM): Besides exergame adherence, the walked distance tracked with the SMM device may have had a significant effect on fall risk reduction. The SMM may have motivated participants to increase their physical activity, especially walking, and consequently this may also have positively affected physiological and psychological components related to fall risk. Some of our participants exceeded recommendations for steps per day (Tudor-Locke, 2002; Tudor-Locke and Bassett, 2004), implying increased motivation to be physically active. In this context, reduced quality of walking associated with frailty is an additional risk factor for falls that should be considered when assessing walking distance in frail older adults (Brodie et al., 2017; Hausdorff et al., 2001). In our sample, participants were rather healthy, which is why we excluded quality of walking for the analysis. Based on the quantitative results, devices like the SMM may represent a valuable extension to indoor exercises such as exergames.

Combination of exergame and SMM: Post-hoc tests indicated that the overall fall risk reduction in the study might have been influenced by the combination of exergame adherence and SMM walked distance. Our results showed that participants who adhered more closely to the exergame and walked more distance by using the SMM reduced their fall risk significantly compared to participants who used only one of the components, and the control group. To the best knowledge of the authors, the literature does not provide statistical analyses on the combined effects of exergames and activity tracking devices. However, our analysis shows that such effects may exist and noticeably contribute to the effectiveness of ICT-based fall prevention systems. Therefore, more research in this area is required to provide more validity about such effects.

6.5.3   Discussion of usage indicators

Several usage indicators, determining the use of ICT-based fall prevention systems by community-dwelling older adults were derived from our qualitative analysis. Those results provide information in order to enhance sustainable use and adherence to such systems by older adults and thereby enable effective ICT-based fall prevention. The following list summarizes the identified usage indicators:

- Older adults should have the possibility of adjusting the level of difficulty (exercise progression) by themselves,
- The ICT-based system should provide simple interaction mechanisms with understandable operating options,
- The ICT-based system should provide a high level of error tolerance with good usability and accessibility,
- New activities (exercises and games) should be implemented to keep the motivation high,
- Multiplayer modes should be included to allow for playing together and/or competition between each other,
- Detailed feedback mechanisms are of utmost importance to allow the older adults to monitor their activities and personal success,
- Creating additional values of such ICT-based systems is recommended (e.g. educational material and advice),
- The ICT-based system should be designed to allow for integrating activities into daily routines of older adults (For instance, social interactions like sports courses outside home),
- Combination of different devices and approaches to support older adults to stay active during the whole day are feasible and recommended.

The application of the SMM device in particular seemed to be a considerable usage indicator for participants. In our qualitative analysis, as suggested above, we were able to identify many factors suggesting that the SMM device may have increased the motivation of participants to take challenging outdoor walks more frequently than they did before the study. Different enablers of motivation were provided for the users by the SMM visualizations in the iStoppFalls system at home. A so-called city comparison, in which the total walked distance of all users for each study site was summed-up and visualized compared to walked distances of other sites was provided and frequently used by the study participants at home. Both the existing literature (Fogg, 2002, 2009) and our qualitative study results suggest that such functionalities seem to amplify older adults' motivation, in terms of self-demanding competition or motivating group challenge.

Most of the identified usage indicators are in line with the findings of previous studies. Gerling et al. (2012) provided design guidelines (e.g. age-inclusive design, dynamic game difficulty and others) which could be confirmed by our study results. Older adults are further motivated to learn and discover new information (detailed feedback, educational material), as shown by results of studies by Lee et al. (2012) and Ballegaard et al. (2008). The referenced literature focused on the design of ICT-systems and/or exergames. Our study extended this approach and included a wearable device, which may have motivated the older adults beyond the usage of an exergame alone. Participants reported to be more motivated to stay active during the day by using the wearable device. This motivational factor was not expected originally, when the SMM was included in the study design, as the device was mainly intended for data collection. It underlines the challenge of such exploratory studies to consider in advance all possible activities and scenarios how participants may use technology at hand. In this context, as Pelle Ehn (1990) stated: *"[...] In any case, a computer artefact is fundamentally what it is to the user in practical use and understanding, not what we find in detached reflection about its properties"* (Ehn, 1990, p. 166). Examination of practice, that is, provides useful additional data.

6.6 Study Limitations

In terms of methodology, we were not able to clearly distinguish the effects of the SMM device from the exergame effect on fall risk reduction. Only the combined effect could be observed, and thus we were not able to discriminate the respective influence. That is due to the study design, which considered the SMM rather as a tool for data collection than a tool for increasing participants' motivation to get physically more active. Regardless, it does seem that there is a synergy in the use of both elements. Further, some older adults who participated in our study might have had a higher motivation to adhere to the study protocol as community-dwelling older adults outside of such a study setting, although we have no evidence to indicate this as yet. It is our view, based on this and other studies we have embarked upon, that such 'study effects' diminish over time. As for the control group, our study design did not allow to track their exercise behavior during the trial. Therefore, we are not able to provide information on the duration or type of exercises control group participants followed. Nonetheless, such information would have been valuable for the assessment of iStoppFalls' effectiveness. With respect to our study sample, most participants had a rather affluent socio-economic background. Therefore, our study cannot provide findings in terms of system effectiveness for older adults with low SES. Subsequently, a drawback in the underlying perspective on SES in this study may be seen in the disregard of participants' income, which generally should be considered when assessing SES. All analyses presented in this paper were not pre-planned and therefore need to be considered as exploratory post-hoc analyses. Finally, the study was not corrected for multiple comparisons, as the sample size was too small. This limitation also applies to effectiveness of iStoppFalls for each subgroup. A similar remark also applies to the statistical power of the results. When comparisons are made between subgroups in this way, we cannot be entirely sure that the levels of statistical significance we discover reflect real effects, although we are confident that they do. It also follows that, except for the derived usage indicators, the results of our study cannot directly be transferred to other ICT-based fall prevention systems and different populations without any restrictions.

6.7   Conclusion

In this paper, we presented results of a study about ICT-based fall prevention in community-dwelling older adults. The analysis focused on the post-hoc analysis of subgroups, which benefit most from an unsupervised ICT-based fall prevention system like iStoppFalls. Using a quantitative approach in addition to qualitative data, the effectiveness of the system for these subgroups was analyzed. Further statistical analyses explained this effectiveness of iStoppFalls for the subgroups of older adults in terms of exergame adherence and SMM usage. Finally, a qualitative analysis revealed indicators, which enabled us to learn more about the system use related to user requirements. Factors that we can describe grossly as 'social' or 'psychological' are evidently important. 'Hygiene' factors include ease of use, error tolerance and appropriate levels of difficulty. Although we have not made comparisons with less older adults, it is possible that such factors are more important for our target population than for the population-at-large. Positive factors seem to have much to do with the availability of social elements, as expressed in the desire for multiplayer games and features, such as city comparisons of group activity performances. Our results suggested that systems like iStoppFalls seem to be most effective for older adults with bad balance and high baseline fall risk, the target group with the highest need. In case of severe frailty of the participants, additional risk factors like quality of walking should be considered for such analyses. Furthermore, our study results in terms of fall risk reduction in older adults showed that the effectiveness of iStoppFalls may depend on indoor home exergames in combination with outdoor activities monitored by wearables such as the SMM. Both activities seemed to reinforce each other in our study. Therefore, the analysis implied that a combination of exergames and activity tracking devices, such as the SMM, is even more effective in reducing fall risk of older adults. Finally, nine indicators that determine the system use of older adults were identified. In combination with the quantitative analysis on the effectiveness of iStoppFalls, these system usage indicators provide a valuable basis for developing effective and acceptable ICT-based fall prevention systems for older adults. Still, there are pending issues concerning cost effectiveness and willingness of older adults to pay for such systems. Therefore, further studies may focus on such aspects to fill this gap regarding exergame-based fall prevention.

7 Exploring User Behaviour to Predict Influencing Factors in Technology use for active and healthy ageing by Older Adults

Abstract

Background and Objectives. Sensor-based health technologies for active and healthy ageing support may improve older adults' health and quality of life. However, people often stop using them before significant impacts on health or quality of life occur, mainly due to insufficient motivation for technology use. Aim of this study was to investigate the influence of technological and social cognitive factors towards the use of sensor-based health technologies by older adults. **Research Design and Methods.** In a mixed methods approach, data was initially obtained from an online questionnaire completed by older health technology users and used in a regression analysis, where factors from the Technology Acceptance Model (TAM) and the Social Cognitive Theory (SCT) served as predictors for health technology use (HTU). Further, in-depth interviews were conducted with older adults to gain insights into technology use and physical activity behavior of older adults. **Results.** The regression analysis showed that the TAM and SCT factors accounted for a significant proportion of variance (39.5%) in HTU. Significant predictors of HTU were physical activity (.399**), social support (.287*), and expectations regarding individual health (.440*) and physical appearance (-.470**), indicating physical activity as mediator for HTU. The qualitative analysis indicated the conflation of technology support with social environments as key for physical activity behavior in older adults. **Discussion and implications.** The findings indicate physical activity as a mediator in HTU by older adults and suggest that the consideration of social factors in health technology design may facilitate the uptake of AHA technologies.

7.1 Introduction

With ongoing technological progress in terms of bandwidth, sensors and data analyses, more and more novel information- and communication technology (ICT)–based solutions edge into the market, providing support for active and healthy ageing (AHA). In order to induce significant health impacts, long-term use of these AHA technologies is necessary. However, health technology use (HTU) for physical activity support is often short-term and therefore anticipated health behavior changes do not happen, especially in the group of older adults (Di Pasquale et al., 2013; Jarman, 2014).

© Springer Fachmedien Wiesbaden GmbH, part of Springer Nature 2018
D. D. Vaziri, *Facilitating Daily Life Integration of Technologies for Active and Healthy Aging*, Informationsmanagement in Theorie und Praxis,
https://doi.org/10.1007/978-3-658-22875-0_7

Previous literature has extensively investigated influencing factors for both, the use of health technologies for active and healthy ageing support and health behavior change at this target group and has identified technology acceptance, intensity, progression, feedback, personal choice and integration into daily life routines as key aspects for sustainable and long-term participation in technology supported physical activity (Ballegaard et al., 2008; Carmichael et al., 2010). Further, the opportunity to network and communicate with peer groups, as well as the provision of goal setting and self-monitoring functionalities was found to be of major importance for older adults to stay involved and follow a technology-based activity program to improve their health and quality of life (Chao et al., 2000; Lee et al., 2012). On the other hand, research on physical activity behavior showed that determinants like health-literacy, self-efficacy, outcome expectations, goals, socio-cultural facilitators and impediments explain most of the variance in physical activity engagement in older adults (Anderson et al., 2006, 2007; Plotnikoff et al., 2013). Such factors have been found to promote motivation in older adults to initiate health behavior change. Even though motivational aspects of physical activity engagement and influencing factors for AHA technology use in older adults have been studied extensively, sustainable and long-term technology supported physical activity engagement remains a major challenge.

For such research questions, either qualitative or quantitative methods are usually applied. A combination of both methods in this context is rarely found. Even though quantitative research provides powerful tools to analyze large sample sizes and identify significant relations between factors influencing HTU, the information extracted by solely quantitative methods may be inadequate to define factors affecting user behavior, for instance user acceptance, self-efficacy or social support (Brannen, 2005). Quantitative methods cannot assess the whole spectrum of heterogeneous experiences of users such as motivational (positive) and 'hygiene' (negative) aspects, which affect their willingness to engage in technology supported physical activity (Herzberg, 1966; Herzberg et al., 2011). However, findings from such quantitative approaches can be taken up by qualitative research to explain user behavior on a more detailed level. We suggest that in addition to quantitative analysis, qualitative research provides a complementary and valuable approach to the understanding of technology supported physical activity engagement by older adults. Therefore, in this study, in a first step, we conducted a regression analysis, based on factors derived from the technology acceptance model (TAM) and the social cognitive theory (SCT) to identify factors influencing technology supported physical activity. For the TAM, we used the original variables proposed by Davis (Davis, 1989), perceived usefulness (PU) and perceived ease of use (PEOU). The influence of both variables on technology acceptance in a healthcare context has been confirmed throughout several studies (Van Schaik et al., 2002; Yi et al., 2006). To the TAM we also added technology experience as an external variable, as it has shown significant effects on technology use in healthcare studies (Ammenwerth et al., 2006; Gagnon et al., 2006). For the SCT, we included the variables proposed by Bandura, namely outcome expectations, self-efficacy, barriers and social support (Bandura, 1986). Studies showed the significant effects of these variables on health behavior (Bandura, 1977, 2004; Heaney, C. A., & Israel, 2008; Williams et al., 2005). Secondly, using a qualitative approach we investigated the factors, which had a significant contribution in the models in more detail, with a focus on physical activity behavior in older adults as it was indicated to be a key mediator for HTU in our study.

The results will provide insights into older adults' motives on physical activity and how these motives may affect their willingness to use technologies for AHA support. Consequently, our study derives design implications that may promote sustainable and long-term technology supported physical activity engagement in this target group.

7.2   Methods

7.2.1   Study design
In total, 203 participants from across Germany participated in this study. In a first quantitative trial, hundred eighty-eight health-technology users completed an anonymized online questionnaire on HTU and physical activity behavior. Medisana, an online platform/company, which distributes a variety of AHA technologies such as activity monitors, pulse oximeters, weight scales, etc., provided access to those participants. Afterwards, fifteen participants took part in a qualitative trial to explore results yielded from the anonymized questionnaires in more detail. Participants for the qualitative trial were recruited by distributing paper-based and electronic ads in newspapers, senior clubs and similar institutions.

7.2.2   Participants
As inclusion criteria, participants who completed the online questionnaire were required to possess at least one out of three sensor-based health technology devices used for physical activity purposes; an activity monitor, a pulse monitor or a pulse oximeter. Participants in the qualitative trial were able to choose at least one of those technologies for a period of two months to explore their usage experiences. Qualified research assistants trained the participants in the qualitative trial to use the technologies appropriately before the study began.

7.2.3 Data collection

Quantitative data collection: An online questionnaire was distributed among participants of the quantitative trial. We integrated three control questions to check the validity of responses. The questionnaire included items on demographic characteristics and physical activity and assessed TAM and SCT constructs with the following validated scales in German language. With respect to the readability of this article, we translated all items into English language.

Perceived usefulness

PU was measured on a 4-item likert scale, ranging from 1-totally disagree to 7-totally agree. The scale was developed and validated by Kothgassner et al. (Kothgassner et al., 2013). The questions were (1) "Using this technology would make things easier", (2) "This technology could support me in completing my daily tasks easier", (3) "If I could afford this technology, I would buy it" and (4) "This technology would support me to complete my daily tasks". The mean value of this construct was formed for each respondent.

Perceived ease of use

Similar to PU, PEOU was measured on a 3-item likert scale, ranging from 1-totally disagree to 7-totally agree (Kothgassner et al., 2013). The questions were (1) "Application of this technology is easy to comprehend", (2) "Application of this technology is easy in general" and (3) "Application of this technology is complicated". Before forming the mean value, the scale of item 3 was inversed, so that a high value for this item would mean the respondent does not find the technology complicated to apply.

Technology experience

In order to determine the respondents' technology experience, we asked for functionalities that they frequently use on their mobile phone. A list of 14 items was provided with the possibility for multiple answers. Each item counted as 1 point. Exemplarily answers were "I use messenger applications like WhatsApp", "I play games", "I write Emails", or "I use the calendar and reminder functions" The sum for each respondent formed the individual technology experience.

Outcome expectations

We measured two dimensions of outcome expectations (health and physical appearance) and therefore applied two scales. Both scales used a 5-point likert scale, ranging from 1-totally disagree to 5-totally agree. The scales were developed by Fuchs (Fuchs, 1994). Each scale begins with an introductory sentence "if I am physically active on a regular basis…" followed by validated items for the specific outcome dimension. For health outcome expectations the items were (1) "…I am more balanced in everyday life", (2) "…I feel good afterwards", (3) "…I gain more self-confidence", (4) "…I become more reliable for everyday-life purposes", (5) "…I do something good for my health", (6) "…My risk of having a heart attack declines" and (7) "…I maintain mobility and flexibility". For physical appearance-related outcome expectations the items were (1) "…this positively affects my looks" and (2) "…I do not have weight problems (any more)". The mean value was formed for each dimension of outcome expectations

Self-efficacy

Self-efficacy with respect to physical activity was measured on a 12-item likert scale, ranging from 1-totally unsure to 7-totally sure. The scale was developed by Fuchs and Schwarzer (Fuchs and Schwarzer, 1994). The scale starts with an introductory sentence "I am sure to be able to be physically active if…" followed by the items (1) "…I am tired", (2) "…I am feeling depressed", (3) "…I am worried", (4) "…I am upset", (5) "…I feel very tense", (6) "…friends are visiting", (7) "…other people want to undertake activities with me", (8) "…my family/partner keeps me busy", (9) "…I find no one to be physically active with", (10) "…the weather is bad", (11) "…I have a lot of work to do" and (12) "…an interesting program is on television". The mean value is formed for each respondent.

Barriers

Barriers to physical activity were measured on a 13-item likert scale, ranging from 1-not at all to 4-very much. The scale was developed by Kraemer and Fuchs (Krämer and Fuchs, 2010). They used some of the items from the existing scale on self-efficacy by Fuchs and Schwarzer (Fuchs and Schwarzer, 1994) and extended it. The scale begins with an introductory question "to what extent do following items impede you to be physically active" followed by the items (1) "I don't feel like it", (2) "I am tired", (3) "I am stressed", (4) "there is still a lot of work to be done", (5) "I am feeling depressed", (6) "I am in a bad mood", (7) "It is cozy at home", (8) "An interesting program is on television", (9) "The weather is bad", (10) "Friends want to undertake activities with me", (11) "I am injured", (12) "I am sick" and (13) "I am in pain". For each respondent the mean value is formed.

Social support

Social support with respect to physical activity was measured on a 5-point likert scale, ranging from 1-almost never to 5-almost always. The scale was developed by Fuchs (Fuchs, 1997). The scale starts with an introductory sentence "People from my circle of friends and acquaintances…" followed by the items (1) "…engage in physical activity with me", (2) "…encourage me to hold onto planned physical activities", (3) "…request me to engage in physical activity together", (4) "…remind me to be physically active and (5) "…support me in organizing my physical activity". The mean values are formed for every respondent.

Physical activity

To collect data on current physical activity levels of participants, we asked them to quantify their hours per week engaged in physical activity. For reasons of clarification, we defined physical activity as any form of activity for example, doing sports and other activities like for example stair ambulation, going for walks, gardening etc.

Technology use
To ascertain data on the use of the regarded AHA technologies, we asked participants if they used the technology devices on a regular basis. In case participants answered "yes" they were required to specify the hours per week they used the AHA technologies.

   Qualitative data collection: To understand older adults' perspectives and motivations for physical activity in more detail, we conducted semi-structured interviews with a focus on physical activity with 15 participants. Interview questions concentrated on participants' motivation for being physically active, the importance of physical activity for them, their individual benefits and risks and their outcome expectations from being physically active. Participants were allowed and required to elaborate freely on these topics. Prior to the semi-structured interviews, we asked participants for their age, technology experience and health status. To assess technology experience, participants were requested to state technologies and functionalities they frequently use; similar to the way we assessed technology experience for the quantitative trial. Health status was not measured objectively, but by asking participants about their current impairments and diseases and how they would describe their physical condition (rather fit or rather impaired). Two trained research assistants conducted and audio-recorded all interviews.

7.2.4   Data analysis
Statistical data analysis: From the initial 188 participants who completed the online questionnaire, we excluded extreme values and outliers, and applied several filters before data analysis commenced: younger than 100 years, reported less than 30 years of education, reported less than 105 hours of physical activity per week, and reported less than 168 hours per week technology use. Participants scoring above the cut-off for one or more of these filters were excluded from the analyses. Furthermore, only participants who reported using at least one of the required devices were selected for the analyses. The required age of participants was 50 years or above. In total, 87 participants were selected for analysis. Data were incomplete for 25 participants, such that analyses were carried out on a total of 62 (11 female) participants (mean age 60.6 ± 8.3), with years of education 13.8 ± 4.3. A linear regression analysis to evaluate the value of technology experience, ease of use, usefulness, self-efficacy, barriers, activity per week, health-related expectations, physical appearance-related expectations and social support in the prediction of

self-reported technology use per week was conducted. All predictors were entered into the regression equation in the same step. Diagnostic tests of tolerance and variance inflation revealed all of the measures fell within acceptable ranges of collinearity (Variance Inflation Factors < 2.5).

Qualitative data analysis: Qualitative data material was analyzed by applying a thematic analysis approach (Braun and Clarke, 2006). Based on transcribed audio files, four coders performed an inductive analysis of the data material and generated main categories. Coding discrepancies were discussed and eliminated by adding, editing or deleting codes, based on the group discussion outcomes. The final code system covered categories relating to the perception of health, the motivation for physical activity, barriers for engaging in physical activity, participants' perceived usefulness and drawbacks of engaging in physical activity and their outcome expectations from engaging in physical activity. Based on the coded data material we derived indicators that encourage participants in our sample to perform physical activity. Those indicators lead to specific implications for the design of sensor-based health technologies, which are presented in the discussion. For the analysis, coders used the software application MAXQDA™ version 12.

7.3   Results

7.3.1   *Quantitative analysis: exploring predictors of health-related technology use*

Descriptive statistics: This section presents the means and standard deviations for participants' scores on technology experience, ease of use, usefulness, self-efficacy, barriers, activity per week, health-related expectations, physical appearance-related expectations, and self-reported technology use and social support.

Tab 6. Descriptive values for the predictor variables used in the regression analysis

Variables	Mean	SD
Technology use (hours)	79,1	69,5
Technology experience	8,7	3,1
Ease of use	4,5	0,8
Usefulness	4,3	1,7
Self-efficacy	4,5	1,2
Barriers	2,1	0,4
Reported physical activity per week (hours)	25,8	22,7
Social support	1,6	0,6
Health-related expectations	4,3	0,7
Physical appearance-related expectations	3,9	0,9

Correlation analysis: A pearson's correlation analysis between technology use and the TAM and SCT variables, which were later added in the regression model was initially applied. In total, HTU showed significant positive (but very low) correlations only with physical activity per week ($r=.322$) and social support ($r=.284$). Interestingly, there was also a negative correlation between HTU and physical appearance-related outcomes.

Regression analysis: Finally, in order to examine the relative contributions of technology experience, ease of use, usefulness, self-efficacy, barriers, activity per week, health-related expectations, physical appearance-related expectations and social support in the prediction of technology use per week, a regression analysis was conducted. All predictor variables were entered in the same step of the analysis, resulting in 39.5% explained variance in self-reported technology use per week. Technology experience, ease of use, usefulness, self-efficacy, and barriers, failed to contribute significantly to the prediction of self-reported technology use per week. Beta weights for the regression equation indicated that physical activity per week ($\beta = .40$, $p < .01$), social support ($\beta = .29$, $p < .05$), health-related expectations ($\beta = .44$, $p < .05$) and physical appearance -related expectations ($\beta = -.47$, $p < .001$) made significant contributions to the prediction of self-reported technology use per week.

*7.3.2   Qualitative analysis: Indicators regarding influencing factors for physi-
cal activity in older adults*

Descriptive statistics: According to the conducted thematic analysis, five different
topics of indicators were revealed that determine older adults' engagement in
physical activity for our sample. This section will describe the results. Table 7
provides an overview of interviewed participants and their characteristics.

Tab 7. Interviewed participant characteristics

ID	Sex	Age	IT-literacy	Physical Status	Marital Status
PN 1	male	74	experienced	fit	lives with partner
PN 2	female	75	experienced	fit	lives alone
PN 3	female	71	experienced	fit	lives with partner
PN 4	male	71	experienced	fit	lives with partner
PN 5	female	78	novice	impaired	lives alone
PN 6	female	81	novice	impaired	lives alone
PN 7	male	72	experienced	fit	lives with partner
PN 9	female	74	experienced	impaired	lives with partner
PN 10	male	78	experienced	fit	lives with partner
PN 11	female	90	novice	impaired	lives with son
PN 12	female	85	experienced	impaired	lives alone
PN 13	female	75	experienced	impaired	lives with partner
PN 14	female	83	novice	fit	lives alone
PN 15	male	68	experienced	fit	lives with partner
PN 16	male	64	experienced	fit	lives with partner

Health improvement: A major topic, participants mentioned with respect
to physical activity engagement, pertained health improvement. One participant
stated that physical activity was mandatory for her to prevent falls and related consequences: *"I need to do that [physical activity] to keep being steady on the legs.
If I fall for instance, I would need a wheel chair. So this [being physically active]*

is very serious for me." (PN 14). Another participant said that physical activity helps her to clear her mind of bad thoughts: *"When I go for a walk, I get rid of my bad thoughts and get new good thoughts. This is why I go for a walk each day at least for an hour [...], this is sport and thinking combined." (PN 2).* Quite similar to the previous quote, one participant elaborated on how physical activity might help her to improve her sleep quality: *"I would do that [physical activity] to improve my emotional well-being. I believe it [physical activity] would help me to follow a more positive daily routine and consequently that might help me to improve my sleep quality." (PN 15).*

Self-determination: Another topic participants were eager to elaborate on was their desire to maintain self-determination. A female participant stated that it is important to engage in physical activity in order to stay independent: *"[...] Well, you will do that [physical activity], especially when you live alone and know that you need to maintain agility and mobility or otherwise stumble into dependency." (PN 8).* A male participant had a similar perspective on physical activity. He was more detailed on his outcome objective and stated that he wanted to continue to work after retirement age: *"My goal is to stay healthy, with respect to physical and cognitive condition. It might be that I resume working in a year. I am 64 now. If so, I could work as much as I want and earn as much money as I want. Therefore, health is my top priority." (PN 4).*

Social participation: Besides health outcomes and self-determination, which seemed important to participants, nearly all participants expressed their desire to maintain capabilities to participate in social life. In general, participants emphasized participation with friends and family. A male participant responded to a question about his motivation to engage in physical activity: *"The view at my children, my grandchildren, and my wife of course. I want to keep participating. That is my motivation for physical activity." (PN 1).* A female participant elaborated on her desire to undertake another trip with her best friends and that they motivate themselves to stay active in order to achieve that goal: *"Doing another trip with my girls. We used to travel together every year, but for 2 years now we couldn't, due to injuries and diseases. Realizing such a trip together again is what drives us to be physically active." (PN 10).* Another participant emphasized that health is a prerequisite to maintain the possibility to interact with social contacts: *"This [social contacts] is of major importance [...]. I live alone but I meet my friends very often. When I imagine that I would not be able to leave the house anymore, this would mean a complete change for me, I would not stay there then. I probably would have to move to a senior home [expresses that he is not eager to do that]. Therefore, social participation is an important aspect with respect to physical and mental health." (PN 8).*

Social support: According to our qualitative results, another important indicator promoting physical activity in older adults seems to be social support. Many interview participants implied that support from their social environment plays a major role when deciding for or against physical activity. A female participant narrated how she motivated a friend to engage in physical activity: *"I had to bolster her. You know she has some health problems and this is why I tried to talk her into regular walking. We now go walking together on a regular basis." (PN 7).* A male participant talked about his family that supports him to be physically active and how this motivates him: *"yes, yes of course. They [family] will know it at first, when I engage in physical activity. I tell them immediately. Their feedback definitely would and could motivate me to be more active." (PN 13).* Another male participant indicated that his wife is very interested in physical activity and for that reason, he learns a bit about it as well: *"It [physical activity] is interesting and you never know what problems and diseases you will face with age. [...], my wife is much more interested in it [physical activity] than me. Thanks to her I get to know it [physical activity] better." (PN 4).*

Being active together: Finally, interview participants implied that the possibility to engage in physical activity with others is an important factor for their motivation. In that context, a female participant mentioned that she does Nordic walking together with her husband: *"I am doing Nordic walking for 15 years now and what I like about it the most is that my husband accompanies me." (PN 2).* Another female participant sees an opportunity in physical activity to interact with and get to know new people: *"Socializing is a result of being physically active, for instance when you have a tennis or golf mate you meet in the morning to play with." (PN 7).* Another participant explains that she would like to do group exercise on a low level: *"I used to be very active you know. I would love to run, but I can't, since it really hurts my muscles. I would love to do group exercise on a low level with other people who also suffer from muscular diseases." (PN 15).* Competing with others seems to be an additional motivator of being active together for some participants. One participant stated: *"I like the competition. Therefore, I visit the gym. I also exercise alone, but I am not so ambitious then, only doing the necessary exercises. When exercising alone it is very easy to find excuses to stop or not even start the training. When I am at the gym I put more effort in my training, also not to lose my face." (PN 7).*

7.4   Discussion

Physical activity can improve health and quality of life in older adults (Gillespie et al., 2012; Mercer et al., 2016; Sherrington et al., 2011). However, a positive effect can only be achieved by means of a sufficient exercise dosage and sustainable training with a good adherence over a longer period of time (Phillips et al.,

2004), and thus the motivation and practices (Wulf et al., 2015c, 2015a) of the participants play an important role. A promising solution to support older adults in this regard are AHA technologies like activity or pulse monitors (Chiauzzi et al., 2015; Gualtieri et al., 2016; Vankipuram et al., 2012). Several studies investigated AHA technology use in older adults and their capability to change older adults' health behavior, for instance towards physical activity. However most studies focused on short-term usage (Bravata et al., 2007; Fausset et al., 2013; Klasnja and Pratt, 2012; McMurdo et al., 2010) and technical or device-related problems (Choe et al., 2014; Li et al., 2010; Rooksby et al., 2014; Sellen and Whittaker, 2010). Despite the number of available concepts like persuasive design (Fogg, 2002, 2009), as well as concepts like embodiment (Merleau-Ponty, 2002; Phillips et al., 2004) and self-efficacy (Bandura, 1977, 1993), the design of AHA technologies for sustainable long-term use in older adults still remains a challenge. Our exploratory study, suggested that HTU in older adults should be disentangled from a technical or device-related perspective, as indicators promoting physical activity engagement in this target group seem to be more relevant predictors for long-term HTU.

7.4.1   Discussion of regression analysis
The regression analysis showed that usefulness, ease-of-use and technology experience were not significant predictors of HTU. This may seem surprising at first since previous research has often underlined the importance of such technology-related factors for HTU (Ammenwerth et al., 2006; Orruño et al., 2011). However, at most of these studies, these factors have been examined 'separately', meaning that they were rarely put in models together with SCT or other factors. According to our results, it may be likely that when considering SCT and TAM factors together, social cognitive factors are more relevant to predict HTU in older adults. A reason might be that older adults perceive AHA technologies merely as tools supporting their transition to a healthy lifestyle and not as tools to support physical activity related goals. Such indications and the fact that the variable "expectations regarding physical appearance" in our regression analysis was negatively associated with HTU, are in line with previous research stating that older adults' motivation to lead a healthy lifestyle and exercise is not to look good, but rather feel good (Reboussin et al., 2000). For that reason, health-related expectations was one of the significant predictors in the regression model. Finally, the important role of social support is confirmed by our findings as well, as it has been found to be a significant predictor for HTU (Scarapicchia et al., 2017).

The most surprising finding of the regression analysis is that self-efficacy did not make a significant contribution to our HTU model, despite it being considered as one of the most important predictors for all kinds of behaviors

(Amireault et al., 2013). However, since physical activity was one of the significant predictors and it is known that self-efficacy is crucial for physical activity, it is likely that this indirect relationship explains this finding.

Research has shown that HTU can be a mediator for physical activity (Graham et al., 2014; Rimmer et al., 2004). Interestingly, our results indicate also the opposite way of this relationship, namely that physical activity, which is known to be influenced by SCT variables like outcome expectations, social support, barriers or self-efficacy (Anderson et al., 2006, 2007; Plotnikoff et al., 2013), may be a mediator for HTU in older adults.

7.4.2   *Implications for the design of AHA technologies*
The quantitative analysis indicated influencing factors for AHA technology use in older adults and suggested physical activity as mediator for HTU. Therefore, subsequently, qualitative interviews conducted with the target group revealed and described factors that promote physical activity behavior in community-dwelling older adults in more detail. Together, these results provide implications that may facilitate the integration of technology supported physical activity in older adults' daily life and thus create opportunities for long-term use. The following list summarizes the derived implications:

- AHA technologies should allow older adults to set individual meta goals. Meta goals should be associated with real life contexts, worth working towards to, for instance upcoming journeys or improved sleep quality.
- AHA technologies should provide support functionalities for friends and family members of older adults. Such functionalities could contain buddy systems, where friends or family members encourage the user to be more active or to achieve set goals. Furthermore, older adults could share results or demand more support from friends and family.
- A certain degree of competition should be encouraged by AHA technologies. For instance, high score rankings to compare with others or daily challenges to motivate older adults to stick to physical activity.
- Social aspects in physical activity are of utmost importance for older adults. AHA technologies should embed their training programs in social contexts, for instance physical activities that end in socializing events like communal cooking or community cafes, where older adults can share their experiences, exchange information and connect with people of the same age or with similar interests.
- AHA technologies should provide functionalities to bring older adults with same physical activity interests together.

Most of the identified indicators are in line with findings of previous studies that investigated promoting factors for physical activity behavior. In their studies, Hall et al. (2010) and Locke and Latham (2002) found that individual goals are associated with greater dedication and are often more valued by older adults (Hall et al., 2010; Locke and Latham, 2002). The positive effects of social support from family and friends on physical activity behavior in older adults has been investigated and confirmed in previous studies (Goldberg and King, 2007; Kahn et al., 2002). Further, social aspects in general play an important role in physical activity behavior of older adults. Studies in this regard proved a significant relation between physical activity and social participation in older adults, suggesting that motivation and adherence to physical activity increases when embedding such activity into social context like social meetings or group exercise (Guedes et al., 2012; Thraen-Borowski et al., 2013). In terms of competition, research on behavior change suggests that older adults are more committed to physical activity and achieving goals, when competing for them with an audience or group (Cialdini, 2001; Locke and Latham, 2002).

The design implications presented in this paper contribute to our understanding of important factors that motivate older adults to engage in physical activity and how technology can support them. Although most of them were mentioned in previous studies, commercially available technologies for AHA support like activity or pulse monitors that aim to promote physical activity in older adults are often used only short-term and do not seem to initiate sustainable health behavior change in that target group (Brodaty et al., 2005; Robinson et al., 2009; Wan et al., 2016). One reason might be that AHA technologies do not sufficiently address differencing motives between younger and older users and their diverging perspectives. Literature indicates that younger adults primarily maintain and improve physical appearances because it is an important component in attracting a partner, whereas older adults exhibit more concern for health outcomes, while being involved with evaluating their lives and searching for meaning (Buhler, 1935; Buhler et al., 1968; Erikson, 1995; Trujillo et al., 2004). Both, our regression analysis and qualitative analysis came to similar results suggesting that socio-cognitive factors like social support, social participation or the social environment in general are stronger involved in motivating older adults to engage in physical activity.

From another perspective, shortcomings in developing technologies for sustainable technology supported physical activity engagement in older adults might be explained by the way we interpret AHA technologies and their purpose for older adults. Most of the commercially available sensor-based health technologies intend to support older adults to be more active by monitoring their activity levels, reminding them to start training sessions, or providing recommendations on how to improve their physical activity behavior. Following technological mediation theories by Ihde, Selinger and Verbeek, the focus of AHA technologies lies mainly on mediating older adults' actions and perceptions with respect to physical activity behavior (Ihde, 1990; Ihde and Selinger, 2003; Verbeek, 2010). Technological artefacts influence how things are revealed to the user, affecting their perceptions and actions, which shape their intention to act. Through this mediation certain perceptions and actions are amplified, while others are reduced (Ihde, 1990; Ihde and Selinger, 2003; Verbeek, 2010). In the context of HTU, AHA technologies aim to amplify for example older adults' capabilities for efficient and effective physical activity and influence them towards a healthy behavior, while at the same time aim to reduce their desires for unhealthy behavior. However, older adults' intentions to engage in physical activity do not necessarily coincide with proposed concepts of available sensor-based health technologies, which primarily lay focus on physical health outcomes. Our study results and scientific literature illustrate that physical health is only one relevant outcome dimension for older adults to engage in physical activity. This dimension is, without questioning, important to older adults and probably is the main reason for them to start being physically active in the first place. Nonetheless, their motivation to continue physical activity long-term decreases quickly, and it seems a major reason is the insufficient consideration of social environments and motives of older adults. In fact, many technologies for AHA support even interfere with social environments and motives of older adults, when pushing them to be more active by causing guilty conscience. Such concepts might work in the beginning, when motivation is still high, but at some point, older adults will feel impeded in their quality of life, as these technologies and concepts do not integrate well in their lives and conflict with their social environment and social desires, when focusing on physical activity and health outcomes only. Therefore, solely concentrating on these aspects may be insufficient, when aiming to promote sustainable technology supported physical activity engagement in older adults. The interactions and interrelations between physical activity, health outcomes and the social environment seem to be of great relevance to this target group.

In conclusion, we argue that already existing AHA technologies such as devices monitoring user behavior and providing context-sensitive feedback, address their functionalities rather well but yet their sole use is not adequate to fill the gap between intention and actual behavior change regarding the initiation and adherence to physical activity programs aiming to improve health in older adults. Therefore, we suggest that there is a need for integrative health platforms, which combine different health devices and adjust their functionalities and health data analyses to social contexts and motives of older adults. Even though, existing literature as well as our study results support the need for integrative and holistic approaches in technology supported physical activity, only few concepts and approaches for integrative health platforms exist (Barnett et al., 2015; Cook et al., 2013; Marcotte et al., 2015). To the best knowledge of the authors, most of these platforms only address health outcomes, do not allow the integration of commercially available sensor-based health technologies and neglect the involvement of older adults' social environments. Integrative health platforms, as we propose them here, will be able to distance from a merely physical health related perspective and consider a broader context, important for sustainable physical activity engagement in older adults, to mediate their actions and perceptions not only with respect to physical activity, but also to an individual and multi-dimensional understanding of health.

7.4.3   *Limitations*

Although this study provides useful information regarding the factors influencing HTU and the importance of physical activity as a mediator, we have to acknowledge several limitations. First, the sample size for the quantitative part of the study is rather small and therefore the results of the regression analysis must be interpreted with caution. However, to our knowledge, this is one of the very few studies that have compared the relative contribution of parameters from the TAM and the SCT together for HTU and therefore the trends, which are to be seen in this data can be used as a base for future research questions. Secondly, as mentioned in the methods section, there were two study arms with different samples, which also differed somewhat in age. Ideally, the same participants who have completed the questionnaire should have been interviewed in order to be able to match better the results of the two approaches. However, we aimed to explore the attitudes of participants who already made use of health-technologies and although doing that via an online platform ensured such, the survey was anonymous and therefore we were not able to further contact these participants and invite them for the additional interview sessions. Lastly, this study did not apply objective assessments, but instead relied only on subjective report of the parameters analyzed (HTU and physical activity). This can potentially introduce bias, especially when applied with older adults, as they are prone to over-/underestimations.

7.5   Conclusion

Much research and resources have been invested in the development of sensor-based health technologies to support AHA such that most of them currently not only provide reliable and valid monitoring but they also have a very sophisticated design, which is attractive to most users. However, it seems that design and functionalities of such devices are only key to users who already are motivated to follow a healthy lifestyle and have integrated such healthy behaviors into their daily routine. People who are on the verge of making a lifestyle change and seek motivation for such a purpose, like for example older adults, use AHA technologies as a support towards this transition. Therefore, integrating AHA technologies into daily life routines and social environments seems to be much more important to older adults than perceiving for instance good usability or physical activity support functionalities. Aspects such as training together with others, pursuing real life goals or involving family and friends into physical activities are just a few examples in this context provided by participants in our study. Technologies for AHA support need to address such differencing desires and motives of older adults. Our results suggest that research and industry should prioritize the design and development of integrative platforms, concepts and services that allow the conflation with social environments of older adults. Therefore, the integration of available AHA technologies into one platform and the alignment of their functionalities to attitudes, practices and motives of older adults may be key for the support of a successful transition towards a healthy lifestyle by means of physical activity. Finally, research should further explore the role of TAM and SCT parameters in HTU as well as the role of physical activity as a mediator in this relationship.

8 Negotiating Contradictions: Engaging Disparate Stakeholder Demands in Designing for Active and Healthy Aging

Abstract
The purpose of this article is to illustrate the complexity and diversity of primary and secondary stakeholder perspectives in the context of health and technology use for active and healthy ageing. Our exploratory interview study will show disparate and similar perspectives of older adults and a range of professional secondary stakeholders like doctors, health insurance agencies or caregivers. We negotiated contradictive perspectives and pointed at crucial challenges for the design of active healthy ageing (AHA) technologies for relevant actors within the healthcare system. Fifteen interviews with older adults and X secondary stakeholders were conducted. Audio-recorded interviews were transcribed and coded using content analysis procedure. In total, four major themes emerged for stakeholder similarities and contradictions, which include perspectives on active and healthy ageing, motivation for prevention, perceived benefits and drawbacks of AHA technologies and concerns with respect to data privacy, control and trust. The insights into and negotiation of stakeholder perspectives have the potential to improve the design of AHA technologies that facilitate their integration into older adults daily lives and thus create opportunities for long-term use.

8.1 Introduction

Life expectancy rates are continuously increasing worldwide due to improvements in public health, nutrition, medicine, education and personal hygiene (Active Ageing, 2002; Barry et al., 2015; Centers for Disease Control and Prevention, 2013). This might endanger the socioeconomic systems of affected countries, as population demographics transit towards an ageing society with only few young citizens and economic costs associated with an ageing population grow. Healthcare professionals face demands to deliver improved care services under ever more constrained budgets. Thus, there is an increasing interest among policy makers and health care professionals alike in approaches and solutions that can help delay the need for healthcare services, prolong independent living, and support older people in what is now termed active healthy ageing (AHA) (Publications Office of the European Union, 2012; Rechel et al., 2013; Vines et al., 2015). Researchers and technology developments have responded to the challenge to support prevention and health in older adults through smart application of technologies (Haluza and Jungwirth, 2015; Stellefson et al., 2015) such as tele-monitoring, remote health

© Springer Fachmedien Wiesbaden GmbH, part of Springer Nature 2018
D. D. Vaziri, *Facilitating Daily Life Integration of Technologies for Active and Healthy Aging*, Informationsmanagement in Theorie und Praxis,
https://doi.org/10.1007/978-3-658-22875-0_8

services or self-monitoring with wearable devices (Publications Office of the European Union, 2012). These modern technologies for AHA support can provide more convenient health promotion compared to old-fashioned programs. Yet many research-based systems oriented towards supporting prevention and health for older adults fail soon after research projects end, which has lead for further needs to improve sustainability and longevity of such systems (Di Pasquale et al., 2013; Jarman, 2014). Part of the problem is that AHA technologies are always deployed into healthcare systems that encompass a variety of secondary stakeholders with contradictory perspectives and goals, for instance with respect to perceptions on well-being, motivation for prevention, data privacy, control and trust. On the other side, success and sustainability of AHA technologies relies on infrastructural requirements like financing models, legal frameworks or social infrastructures. Thus for a system to be sustainable requires not only a good technological design and willing users but also support from physicians, policy makers and local governments for integrating such new technologies into the extant systems of health and oversight.

Approaches such as user-centered and participatory design (PD) already suggest involving primary end-users and secondary stakeholders in the design and development of related AHA technologies (Pilemalm and Timpka, 2008; van Gemert-Pijnen et al., 2011). Nonetheless, current HCI literature on the design and development of technologies for AHA rarely considers primary and secondary stakeholder perspectives at the same time as a major part of the design of respective ICT-based solutions. We argue that development of technologies promoting health and prevention in older people hinges on ensuring that all relevant stakeholders are allowed to act together. Here we present results from an exploratory interview study conducted with primary and a range of secondary stakeholders. Our objective was to identify, analyze and negotiate perspectives of end-users and stakeholders relevant for the design of AHA technologies. The results illustrate important contradictions, commonalities and interactions in end-user and stakeholder perspectives with respect to AHA technology use and suggests how to negotiate them for a design of AHA technologies with sustainable impact on the corresponding healthcare system.

8.2   Related work

8.2.1   Policies and ageing society

The European commission regularly announces research calls, under the name of Active and Healthy Ageing (AHA), to tackle the challenges of an ageing society with innovative health care products and services for older adults. Increasingly, projects focus the development of ICT-based interventions for the support of AHA in older adults. Such projects are expected to yield solutions that promote sustainability in AHA in the form of long-term use, integration into daily routines and effectiveness in improving health. In this context, ICT-based solutions for AHA must address a shift in the power relationships within healthcare systems, away from the unrestrained authority of the medical professional and towards a more collaborative partnership with patients taking on a greater responsibility and a more active role in managing their own wellbeing. To manage such responsibilities, users need not only to understand the possibilities of eHealth tools but they also need to feel that they have control over how they interact with them (Publications Office of the European Union, 2012). A better understanding of the possibilities that modern technologies provide in terms of AHA requires education in health literacy (Sørensen et al., 2012). From a policy perspective, a high level of health literacy is considered of substantial benefit for maintaining health (Sørensen et al., 2012; Wolf et al., 2005). At the same time, the empowerment of older adults with control over technology and their own health data is required to address the desire to be treated as equals rather than passively monitored subjects (Mihailidis et al., 2008; Steele et al., 2009).

Despite significant efforts on the part of policy and research in the development of AHA technologies, we often see that in real world contexts primary users lose motivation quickly and stop using AHA technologies. This happens most often because new health systems fail to successfully integrate into daily routines of older adults, including interactions with other stakeholders (Ogonowski et al., 2016). Policy and research might need to think broader and start to consider the healthcare system as a space where all stakeholders, not only primary end users like older adults, but also secondary stakeholders like physicians, health insurance companies or policy makers interact and influence one another and affect the sustainability and long-term use of technologies for AHA support.

8.2.2   AHA technologies for older adults

Technology can be a very valuable tool for addressing the rising health needs of the aging population. It has the potential to improve the situation of older adults as well as their family and/or care-givers (Billipp, 2001) by monitoring performance and behavior, enabling physical and cognitive exercising at home and also promoting social networks. Research on the use of ICTs has shown that it may positively impact quality of life for older adults (Jacqueline K. Eastman and Rajesh Iyer, 2004; Schulz et al., 2015) by improving social support and psycho-social well-being (Charness and Schaie, 2003; White et al., 2002). Internet access appears to improve connections with the outside world and may help older adults avoid or reduce feelings of social isolation and loneliness (Bradley and Poppen, 2003; White et al., 1999).

Wearable sensors, mobile devices and exergames have also been used in order to detect early-risk and assess frailty in different domains of ageing, like physical activity, cognition, emotional state and social connectedness and have provided preventive measures for such domains (Botella et al., 2012; Gschwind et al., 2015; Schoene et al., 2015). There is strong evidence for the effectiveness of technology-based interventions for the promotion of health (Fanning et al., 2012; Gschwind et al., 2015; Vaziri et al., 2017). However, a major challenge is motivating long-term use of such technologies. Here, major challenges for the design of AHA technologies for older adults are for instance, (1) older adults' limited capability to understand technical terms or artefacts and articulate requirements and obligations, (2) their longer learning curve and need for multiple iterations to get used to AHA technologies and (3) their need for social support infrastructures not only for technical issues but for social participation (Chaudhry et al., 2016; Eisma et al., 2003; Lindsay et al., 2012). Moreover, privacy and trust play an important role as many current technical solutions collect personal health data in order to provide new opportunities for connecting end users with secondary stakeholders such as doctors, business companies, health insurance companies or the government (Braun, 2013; Heart and Kalderon, 2013; Lee et al., 2013; Miller and Bell, 2012; Morris and Venkatesh, 2000). ICT interventions need to address such considerations and should be personalized and tailored to the needs of the end users (Andrews and Williams, 2014; Kiosses et al., 2011). However, the design of AHA technologies should not only align to the needs of primary end users but must also take into account the perspectives of secondary stakeholders such as physicians or health insurance companies. System design must also address healthcare system challenges such as sustainability, access and equity, which are framed by policies and strongly influence technology use in the healthcare system (Coddington, 2000; Giacomini et al., 2007; Lehoux et al., 2014). Such challenges for ICT design require appropriate methodologies and instruments that allow for collaboration and cooperation of all relevant stakeholders as well as mediation and moderation between them by designers.

8.2.3   *Engaging stakeholders in the design for AHA*

In recent years, a considerable amount of research in the area of designing technologies for AHA has been implemented widely. A large part of that research concentrates on primary stakeholders, examining how technology can motivate people to adapt to a healthy lifestyle or how people interact with AHA technologies and integrate them into their daily routines (Bisafar and Parker, 2016; Cila et al., 2016; Schaefbauer et al., 2015). While researchers have considered secondary stakeholders extensively, this work has mainly focused on technology use in professional, more clinical or care related, healthcare contexts (Fitzpatrick and El-lingsen, 2013). Further, very little research has considered the integration of primary and secondary stakeholder perspectives for the design of AHA technologies for private end users, even though it stands to reason that both stakeholder groups have divergent interests that might affect technology design. Jacobs et al. (2015) compared health information sharing preferences between breast cancer patients, doctors and navigators. They found discrepancies between stakeholder perspectives, such as the hesitation to share emotional issues like loneliness or satisfaction with care (Jacobs et al., 2015). Gerling et al. (2015) investigated long-term use of motion-based video games in care home settings. They found that the integration of caregiver staff is important for the upkeep of technology by older adults in this setting. However, perspectives of caregiver staff on such technologies remained unaddressed (Gerling et al., 2015). Latulipe et al. (2015) investigated technology use of a patient portal by low-income patients of health care centers and professional caregivers. From their study, they derived design considerations like providing system use training, audio lexicons of medical reports or compilations of advice and encouragement messages to make the use of the patient portal for low-income older adults and caregivers more palatable (Latulipe et al., 2015).

To design technologies for AHA support that promote long-term use in older adults, the inclusion and negotiation of all relevant stakeholder perspectives like older adults, doctors, policy makers or health insurance companies is needed. A major challenge when engaging all stakeholders lies in contradictory perspectives and their interactions regarding health and technology use. Addressing all perspectives and demands is impossible. Therefore, designers have to face trade-offs and make decisions how to deal with them. PD provides some solutions to such challenges and offers a framework for collaboration and cooperation among primary end users, researchers and relevant stakeholders.

8.3   Methods

8.3.1   Research setting and data collection
The study is part of the European research project <project name> with the goal to design a technology-based health platform for primary end users and secondary stakeholders. The platform will combine various health devices and software applications that support older people in AHA, for instance by monitoring their activity or nutrition, and allow cooperation and collaboration with doctors for example via remote health services, like telemedicine. From previous projects, we learned that the success of such AHA technologies requires a thorough consideration of both, primary and secondary stakeholders (Andersen et al., 2011; Chen et al., 2013; Ogonowski et al., 2016). Therefore, we contacted both primary end users

and secondary stakeholders via telephone and email and sent inquiries for interviews, to learn about their perspectives and opinions on AHA and technology use. The interviews encompassed questions in the areas of prevention, health and technology. In total, we conducted 26 interviews with primary end users and secondary stakeholders. Fifteen primary end users with an average age of 76 years were interviewed. The primary end user sample involved nine female and six male participants with a balanced ratio of physically fit and physically impaired health. In order to assess end users' IT-literacy, we asked them for their experiences with modern technologies such as smartphones, personal computers or tablet-PC's. In case participants used at least one of these technologies on a regular basis, we classified them as experienced. Otherwise, they were classified as novice (11 experienced and 4 novices). As an indicator for potential social isolation, we asked participants whether they live together with someone or live alone. In total, ten participants lived together with someone, and five participants lived alone. Table 8 provides an overview.

Tab 8. Sample characteristics

ID	Sex	Age	IT-literacy	Physical Status	Marital Status
PN 1	male	74	experienced	fit	lives with partner
PN 2	female	75	experienced	fit	lives alone
PN 3	female	71	experienced	fit	lives with partner
PN 4	male	71	experienced	fit	lives with partner
PN 5	female	78	novice	impaired	lives alone
PN 6	female	81	novice	impaired	lives alone
PN 7	male	72	experienced	fit	lives with partner
PN 9	female	74	experienced	impaired	lives with partner
PN 10	male	78	experienced	fit	lives with partner
PN 11	female	90	novice	impaired	lives with son
PN 12	female	85	experienced	impaired	lives alone
PN 13	female	75	experienced	impaired	lives with partner
PN 14	female	83	novice	fit	lives alone
PN 15	male	68	experienced	fit	lives with partner
PN 16	male	64	experienced	fit	lives with partner

Interviews with primary end users were conducted in their homes to provide a familiar non-artificial situation. The interviews had an average duration of 50 minutes, following a semi-structured guideline and were performed by two researchers. One researcher led the interview, while the other researcher took notes and observed. All interviews were audio recorded. As exemplarily questions, we asked participants how much they value a healthy lifestyle, what motivates them to undertake preventive measures, how much they value social contacts and what concerns and opportunities they see in technology use and health data storage.

We conducted eleven interviews with representatives of seven different types of organizations within the healthcare system. Our sample included one business company, one health insurance company (HIC), two policy makers, three physiotherapists, one NGO, two doctors and one caregiver. We conducted semi-structured interviews with secondary stakeholders at the location of their choosing, typically at their organization. Interviews were performed by two researchers and lasted on average about 50 minutes. All interviews with secondary stakeholders were audio recorded. For example, we asked secondary stakeholders how important health prevention in older adults is and how it can be improved, how they motivate older adults to act healthier and what concerns and opportunities they see in technology use and health data storage for professional health services.

8.3.2   Data analysis approach

The qualitative data material from audio records and text notes, collected in observations, was analyzed by applying a thematic analysis approach (Braun and Clarke, 2006). Trained research associates transcribed all audio-recorded interviews. Based on the transcripts, four coders performed an inductive analysis of the data material and generated main categories. Coding discrepancies were discussed and eliminated by adding, editing or deleting codes, based on the group discussion outcomes. The final code system covered categories relating to the perception of health, the motivation for prevention initiatives, perceived usefulness and drawbacks of AHA technologies, data privacy, trust and control over data and technologies. Based on the coded data material we derived different primary end user and secondary stakeholder perspectives on prevention, health and technology. For the analysis, coders used the software application MAXQDA [54].

8.4   Findings

Our data analysis revealed both many similarities and striking discrepancies in attitudes and expectations between primary and secondary stakeholders. Given the diversity of the secondary stakeholders in this study, their attitudes towards many aspects of AHA were not uniform either. Below we discuss these agreements and disagreements across four themes addressing the basic conception of what entails AHA, expectations for motivation among older adults, discussions of how useful technologies for AHA support might be and their drawbacks and, finally, considerations of privacy and data disclosure. We also consider how the relevant disagreements might be negotiated with a view to designing an actual AHA system.

8.4.1   Perspectives on active and healthy ageing

While perspectives on what entails Active and Healthy Aging vary, participants in our study consistently highlighted four components: (1) social participation, (2) physical activity, (3) nutrition and (4) sleep. Both primary and secondary stakeholders shared common perspectives on social participation and physical activity, but their perspectives on nutrition and sleep were rather different.

Agreements on physical activity & social participation

In our study, we learned that older people and secondary stakeholders understand AHA primarily in light of physical activity and social participation. Older people understand AHA as a convenient means to maintain social contacts and they would welcome technology to support this activity: *"Well, I need social contacts! [...] This is why I want to stay in my own home as long as possible, and of course, I appreciate any technology that can help me with staying healthy longer [...]." (PN 12, female, 85 years).* Further, some older people see a distinct association between social participation and physical activity that motivates them in AHA. *"[...] I like to go to the gym. But, when you go alone all the time you easily find excuses not to go. And such excuses will become more and more. That is why I prefer to go to the gym with others, also to compete and put more effort. It is a psychological effect." (PN 12, female, 85 years).* Secondary stakeholders share these perspectives and emphasize the importance of social participation and physical activity from a professional point of view: *"Physical activity and social participation have a strong correlation. If someone lives alone at home, you can see that he takes considerably less care of his physical health. [...]" (Doctor).*

Disagreements on sleep and nutrition
Both the older people and secondary stakeholders understand nutrition as an important factor for AHA. However, their perspectives on what healthy nutrition ought to look like deviate considerably. For instance, older people mentioned how important it was for them to enjoy food. For many eating needed to remain a pleasurable experience where the demands of healthy nutrition were perceived in conflict with this desire: *"Of course, healthy nutrition is important to me. However, it clearly contradicts my lust for food [...]." (PN 7, male, 72 years).* Additionally, some remarked that they do not appreciate suggestions and recommendations regarding food and nutrition, as they feel patronized by that: *"Yes, it [healthy nutrition] has a specific importance. However, it is not so important to me that I would allow for someone to tell me what to eat and what not. I don't like to be patronized" (PN 1, male, 74 years).* This stands in stark contrast to perspectives of secondary stakeholders, who shared the common opinion that older people lack the physical and cognitive capabilities to eat in a healthy manner and require more guidance and support in healthy cooking and nutrition: *"Yes indeed, [...] with age capabilities to cook properly every day decrease. We could do more here. For example, collaborative cooking communities or support and guidance from professionals." (Caregiver).* Thus while the secondary stakeholders insisted that nutrition must be a focus for system design, the primary users expressed a great deal of caution about this aspect of the system.

Sleep was another aspect of daily life that generated a diversity of opinions both between and among the stakeholder groups. About half of our older participants did not see sleep as an important parameter for AHA: *"well, that [sleep deficit] does not impair me in my daily life. If I slept not enough, I simply catch up on some sleep over the day."* (PN 2, female, 75 years). For the other half of older people, sleep was a factor they associate with personal well-being: *"If I really managed to sleep through the night, I am a totally different person and happier the next day."* (PN 13, female, 75 years). In contrast, secondary stakeholders have very consistent perspectives on the positive effects of sleep for AHA: *"[...] there is correlation between well-being and sleep. A healthy sleep is crucial for the life quality of older people and the status of sleep. I believe that sleep deficit is a very important topic and that is currently not considered sufficiently."* (Caregiver). Once again, we see that while there are clear and important factors for AHA from a clinical perspective, the experiential nature of these aspects of daily life for the primary users often put the stakeholder groups in conflict.

8.4.2   Motivation for prevention activities

For prevention measures to succeed, motivation is key. Motivation for disease prevention, especially in older people, is a highly individual and situational multidimensional construct. In our study, we found that primary end users and secondary stakeholders shared a common understanding that family, social participation, awareness and independence are important factors for motivation in the context of AHA. The major disagreements primarily emerged in the discussions of financial incentives and structures for implementation of AHA.

Agreements on the importance of family and friends

The desire to maintain the ability to participate with family and relatives drives many older people to perform disease prevention activities: *"I've got a lot [of motivation]. First and foremost, my granddaughter [...]. I want to see how my granddaughter turns out. [...]. That's my motivation, my encouragement."* (PN 11, female, 90 years). This was especially evident for participants who lived with family and enjoyed a larger family and social circles: *"[...] top priority are my kids, my grandchildren, my wife of course. I want to be able to participate and keep up with them."* (PN 10, male, 78 years). Secondary stakeholders also understood that familial connections can act as crucial enablers for disease prevention in older people and that prevention programs should address this topic more intensely. *"[...] when children enter the field, you have a different perspective on health. [...]. It is easier to motivate families for prevention than people who live alone [...]."* (HIC).

Alongside family, strong social relationships and social participation are convincing criteria from the perspective of older people in our sample to follow prevention measures: *"[...] social contacts are the most important thing. With age you learn to appreciate this. For instance, I live alone. If I lose my mobility and would not be able to leave the house anymore to meet with friends, I would want to move to a care home facility. [...]. It [social contacts] is a very important aspect for health."* (PN 2, female, 75 years). This became especially important for participants who lived alone and did not have much family nearby. Here participation in organized social activities gained in importance: *"[...] by playing bridge [...] you are not alone, you belong to a community, and you have company for lunch and dinner. You don't have to go to your room and watch TV alone. It is much more enjoyable to do things together than sitting in your room alone watching TV."* (PN 12, female, 85 years). Secondary stakeholders also emphasized the risk of social isolation and the importance of social participation in older people. They understood social participation as therapy and potential cure for health problems in older people: *"Human contacts are a form of therapy. Formerly you had that for example in cabs or at the hairdresser. This has gone in today's society, which has become very cold. Health issues are strongly intermeshed with that [social contacts]."* (Physiotherapist).

The promotion of awareness and the consequences that can follow from an unhealthy lifestyle seem to be valid factors for primary end users and secondary stakeholders, determining another form of motivation for practicing health prevention activities. Older people are aware of family members or relatives who followed an unhealthy lifestyle and had to deal with the consequences: *"There are bad examples [for an unhealthy lifestyle]. My father died very young. [...]. He is my motivator for acting more healthy."* (PN 15, male, 68 years). Accordingly, many secondary stakeholders discussed how such personal examples could become effective motivators: *"You motivate older people by making it very clear to them that they will have to move to a care home facility when they lose their mobility and that life quality is much higher with sufficient mobility."* (Doctor).

Most of the primary stakeholder concerns about maintaining social ties and participating in social events hinged on the ability to remain independent. This then became the single biggest motivator for health promotion: *"The reason I take care of my health is that I want to be able stand on my legs independently and upright and walk wherever I want."* (PN 10, male, 78 years). From the secondary stakeholder point of view, independence is crucial in health promotion and in reduction of health costs overall as these two factors are seen as interdependent: *"With increasing dependency, motivation for health prevention decreases and therefore it will be much harder to take influence on the motivation of older people."* (Caregiver).

Disagreements on financial incentives as motivators

A prominent factor in our data was financial incentives and constraints associated with health. Monetary incentives are a common method in health promotion. However, they are rarely discussed as influencing factors in the design process of AHA technologies. Our analysis showed that primary end users associate wealth with personal health: *"Maintain healthy, physically and mentally. It might be that I return to work soon. [...]. I can continue to work as much as I want and earn as much money as I want. I need to be healthy for that. Health is the most important thing." (PN 16, male, 64 years).*

There was significant diversity of perspectives among caregivers, physiotherapists and policy makers regarding monetary incentives to promote health in older people. For instance, caregivers and physiotherapists have very contradictive perspectives on monetary incentives. Caregivers were willing to use rewards for older people for acting healthy: *"[...] it is a score system, where you can collect 100 points. Points can be collected by doing different activities in different topics such as mobility, strength, cognition, endurance and enjoyment of life. Once a person collected 100 points he will be rewarded with a 10 € voucher, valid for the city gallery [...]. (Caregiver).* On the contrary, physiotherapists would vouch for not offering health interventions free of charge, so that they become of value for older people. *"[...] If you offer something [prevention measure] that is entirely free of charge, it is not as important to someone as when you paid money for it." (Physiotherapist).* Policy makers' perspective on this matter pinpoints the fact that older people often lack access to preventive measures and initiatives, for instance, due to insufficient funds. *"[...] I might need to reduce barriers first, like the fact that as an older person I maybe cannot afford participating in a sports club. He might put his existential fears first. It is our job as society to find alternative ways to lead older people into prevention measures" (Policy maker).* From their perspective, it is rather more important to find alternative ways to engage and motivate socially disadvantaged people for prevention measures than offer highly complex approaches that fulfill newest standards for the middle class: *"[...] develop prevention activities in the context of social integration for instance. Make them free of charge, offer them in quartiers. That would increase the motivation of these [socially disadvantaged] people more sustainably than offering new fancy prevention methods for middle class elderlies, who probably have a higher motivation to test such programs as they can afford them." (Policy maker).* Here we see that design decisions might differ significantly depending on the stakeholders involved. Clearly, the secondary stakeholders that engaged with older adults directly, such as caregivers or physiotherapists, thought of financial incentives as motivational factors that could be directly manipulated. Policy makers in contrast were concerned with bigger structural issues such as access to healthcare. Yet from a design point of view, these perspectives would have to be negotiated carefully. After all, the very meaning of financial incentives might change significantly depending on the financial structures in place to enable the use of the AHA system in the first place.

8.4.3   Benefits and drawbacks of AHA technology use

Potential long-term use of AHA technologies requires that technological solutions address benefits and usefulness from all stakeholder perspectives very explicitly. Our study revealed that both, primary end users and secondary stakeholders agree upon three benefits and two drawbacks of AHA technology use. They believe that monitoring health-related objectives like physical activity, improved security like obstacle warnings and effort reduction as in reduced necessity to visit doctors are major benefits. On the other hand, primary end users and secondary stakeholders fear that technology use will increase the risk of dependency in terms of overreliance and impair older peoples' self-awareness, as they might unlearn to listen to natural body signals. Further, we were able to reveal that primary end users and secondary stakeholders disagree on the advantages of improved health literacy in older people and the benefits, technology might provide for communication between older people and professionals.

Agreements on benefits
Many primary end users valued the possibility to self-track specific aspects of the body, such as sleep or breathing: *"Since I use this health device, I tell myself, 'Let's go check upon yourself'. I take a look at my sleep behavior or how I breathed. [...]:"* (PN 12, female, 85 years). The benefits for some participants lie in the comfort and convenience of digital tracking of things such as measuring walking distance with electronic pedometers compared to old fashioned paper and pencil solutions: *"I think, yes. In the long run I would use it. After all, I would monitor myself on a regular basis, like seeing how many steps I have taken and so on. Keeping track of such things would be quite comfortable with technology."* (PN 16, male, 64 years). The heath data produced by activity tracking also clearly represented a host of opportunities for many secondary stakeholders: *"[...] I think it can be very useful, measuring the pulse, blood pressure, etc. For instance, I would see whether the patient achieved the set distance goal or I need to encourage him more, with individual measurements [...]."* (Doctor).

Another important factor for usefulness of AHA technologies primary end users and secondary stakeholders agreed on is security. Some older people see benefit in technology use, as it might prevent them from physical harm: *"If I would have a device that warns me when I should pay attention to my environment. That would be great!" (PN 5, female, 78 years).* Other older participants saw in technology a useful minder that could help them to find their way home or tell other people their location to pick them up: *"With age, disorders become more frequent, orientation for instance […]. I would find it useful if relatives could find my current location, in case I lose my orientation." (PN 10, male, 78 years).* Secondary stakeholders also agreed that technology can prevent physical harm and provide support for disorientation and helplessness. *"Technology might be useful in situations where they cannot act self-determined anymore or when they lost orientation and wander around." (NGO).*

Finally, primary end users see a major benefit of AHA technologies in the potential to reduce health-related efforts like going to the doctor or hospital less frequently. *"[…] that [health measurements collected by technology] may be very useful, so that you don't need go to the doctor often […]." (PN 2, female, 75 years).* Secondary stakeholders agree and welcome technology use for these purposes: *"[…] generally it [working over the internet] is very convenient. For many things you don't need patients to be on-site. Especially with older people, many things can be done via Skype for instance." (Physiotherapist).* Thus, AHA systems were seen as a way to gain more support in day-to-day functioning and to reduce interactions with the health system – a clear benefit seen from all stakeholders.

Agreements on drawbacks

Yet acceptance of technology was not universal and with good reason. Many of the older participants were cautious in using AHA technologies, as they feared this could lead to being controlled by technologies, institutions or other people and to loss of independence. *"No! For God's sake, that's terrible. I already said I don't want to live a life that is controlled by other people. I don't want to be ruled by equipment and apparatus. I think you've got to be very careful […]. Because we are used to living independently" (PN 2, female, 75 years).* Given the clear potential of personal health data collection and the sheer scale of access to the intimate daily functioning of older adults, secondary stakeholders agree that data leakage and even unintentional control could be a risk and wished for clear limits to avoid external control by institutions or other people: *"There have to be ethical limits. No one wants to be externally controlled. I am very concerned about that." (NGO).*

Alongside concerns of external control, many of our older participants were also concerned with potential impairment of self-awareness, as they would rely too much on the data and recommendations provided by the technologies. Health systems can produce convenient data about sleep or walking but many crucial activities would of course remain unmeasured and potentially could become deprioritized: *[…] I would have a problem with controlling myself all the time. I would be concerned with data about my health most of the time. It could irritate me easily, for instance if it tells me my pulse is too high and I would not know what it means or what to do. [...]." (PN 4, male, 71 years).* Despite the excitement around potential for health data, secondary stakeholders also shared the concern that older people could focus too much on technology and stop listening to natural signals of their body: *"[...] you know what I mean. If in the end, my self-awareness suffers, because I only can tell how I feel by looking at the display. Something is wrong then. It is a very thin line." (Physiotherapist).*

Disagreements on drawbacks
A major factor for older end users to use AHA technologies is the possibility to improve their health literacy: *"[...] I would use technology to gather more information on healthy lifestyle and health in general." (PN 10, male, 78 years).* Some thought this could also strengthen their position in their interactions with health professionals: *"Yes, maybe this technology could support me if I need to assert myself at the doctor's or at some other organization. I'd be able to say we haven't taken this and that into account and maybe we could try such-and-such to get to the bottom of the problem." (PN 13, female, 75 years).*

Despite the support for health literacy clearly expressed in recent EU policy statements, there was considerable disagreement from the secondary stakeholders about the value of increased health literacy. For example, doctors and physiotherapists feared an increased resistance to advice in older people, which would aggravate professional health services: *"It is like you ask Dr. Google all the time. It is not good, if people rely on their apps or the information in the internet more than on their doctor's advice." (Physiotherapist).* The policy makers were less negative, but also acknowledged this issue: *"In general it is the right approach to increase health literacy in older people. However, we should prevent situations where the patient thinks he knows better than the professional does" (Policy maker).*

Yet perhaps the strongest disagreement of all we discovered around the most common of IT system features – reminders. Secondary stakeholders anticipated improved communication with professionals and patients and were keen on the possibility to send reminders to their patients: *"Interconnection is important! For instance, if I want to share information with my team or patients, I could create a virtual communication space. I would be able to remind them frequently, like five times a day to take care of things." (Physiotherapist).* In contrast, primary end users explicitly mentioned during our interviews that they are not fond of such reminders and notifications. In fact, the reaction was often so negative that it became a barrier for using such technologies: *"I find it extremely annoying, when I get notifications or recommendations what to eat, when I should stop to eat or when I should adapt my nutrition, or when it tells me I gained a bit of weight. I know such things myself. [...]." (PN 15, male, 68 years).* Primary end users acknowledged the dangers of failing memory but constant reminders were strongly perceived as problematic loss of independence.

8.4.4   Data privacy, control and trust

Data privacy, personal control and trust are crucial considerations for the design of AHA health systems given the prominence of personal health data and the flow of this data between primary and secondary stakeholders. Primary and secondary stakeholders had similar perspectives on data privacy. However, their perspectives differed in terms of trust in institutions and control over personal health data. In fact, these perspectives were so divergent among primary and secondary stakeholders that we could not simply align them into agreements and disagreements. Instead, what we found was that concerns about data privacy and trust in the context of technology use for AHA build upon pre-existing relationships between primary end users and secondary stakeholders and that the desire for data control derives from this. These relationships can at times require negotiations of new accountabilities when technology use for AHA support enters the field.

All primary end users in our study had major concerns in sharing their health data with HIC's, mainly because they feared increased health insurance contributions: *"When you talk about health related data, data privacy becomes much more important compared to other data. Sharing health related data with e.g. health insurance agencies can be harmful for yourself" (PN 15, male, 68 years).* Noticeably, many secondary stakeholders shared these concerns: *"I think it is very important, there is a high risk for data abuse and increased health insurance contributions when sharing health data with health insurance companies" (Doctor).* Such concerns may arise due to pre-existing and somewhat adversarial relationships primary end users and secondary stakeholders like doctors or physicians have with HIC's. While end users, doctors and physicians are reliant on subsidies and payments for healthcare products and services, HIC's need to be economical and minimize expenditures. These goals are often in conflict and data sharing can be seen as too invasive. Besides, many of the fears and concerns about data privacy are also fueled by and derived from the media: *"If you look at the media in the last two years, you see how private data is being abused and where you can buy all that data! [...]" (Policy maker).* Having said this, trust of many older people and secondary stakeholders in health institutions is corrupted and relationships with HIC's therefore have no real foundation of trust.

Technologies for AHA support need to build upon these pre-existing relationships and perspectives and can aggravate data privacy concerns and mistrust even more, as older people, and in many cases also secondary stakeholders, lack

IT-literacy to understand how technology works and how health data is being processed: *"I am quite sensitive in that area [personal data]. I barely upload personal data in the internet, for the reason I do not know and do not understand who uses the data and what happens with them." (PN 1, male, 74 years).* Despite the admitted lack of literacy, many end-users demanded control over their health data disclosure: *"[data] transfer? Yes, but under the condition that I will be asked for permission up front. The owner of that data should decide whether they may be transferred or not. I expect that to be sorted out before I use such technologies." (PN 4, male, 71 years).* This need was keenly understood by many of the secondary stakeholders that argued for increased transparency of health data processing: *"[To trust in technology] it would take very crystal clear, transparent structures on how data is being handled." (Caregiver).* Full control over health data and determination about who can see or use that data are key factors for both stakeholder groups for long-term AHA technology use. This sentiment is shared by prior research and current policy reports (Latulipe et al., 2015; Publications Office of the European Union, 2012; Schulz et al., 2015). Policy makers expressed significant support for enabling older people with more control over their health data: *"[...] and people should have the possibility to inform themselves and come to a decision, whether they want that [sharing health data] or not." (Policy maker).* By doing so, policy intends to build a foundation for long-term use of technologies for AHA support by older people. Our participants illustrated clearly the need for such mechanisms, but still policy needs to tackle broader issues on an infrastructural level to promote sustainable impacts of AHA technologies on the healthcare system. Promoting long-term use by older people through means of health data control is only one part of it.

8.5   Discussion

Our results show that there are major contradictions across various issues. However, it seems that a bigger set of problems are the fundamentally different conceptions of central notions such as independence and well-being among both primary and secondary stakeholders. While everybody agrees on the importance of independence and well-being, many of the disagreements, presented in this paper, stem from the different meanings of both concepts for older people and secondary stakeholders. As most of the current literature and research on the design and development of technologies for AHA support concentrates on either primary end users or secondary stakeholders (Fritz et al., 2014; Gerling et al., 2015; Schorch et al., 2016; Uzor and Baillie, 2014), exposure of such deceptive agreements and different understandings is critical. In what follows, we discuss deceptive agreements and the sources of disagreements around independence and well-being. We then consider how these can be negotiated for the design of AHA technologies and what political and infrastructural requirements are necessary for the sustainable deployment of such technologies into the healthcare system.

8.5.1   Conceptions of independence

Our data demonstrate that primary and secondary stakeholders are in significant agreement about the importance of supporting independence as a major motivating factor for health promotion and disease prevention activities. We then detailed significant disagreement about factors such as health-literacy and technology design functionalities such as reminders both seen as key to maintaining independence but by different stakeholders. These disagreements, we argue, stem from different conceptions of the notion of independence, making the stated agreement on the subject deceptive.

Health Literacy: Improvements in health literacy are supported as a matter of EU policy and this policy seems to align with the attitudes of older adults in our study. For older adults, health literacy was important to their sense of independence. They wished to understand check-up procedures and their outcomes, and to be able to demand alternative procedures, in case of doubt. Health literacy in this context was seen as a way to shift from passive receipt of medical instruction to active involvement in own health. As Lorenzen-Huber et al (2011) found, older people do not want to be passively monitored subjects, but desire to be treated as equals (Lorenzen-Huber et al., 2011).

This was cause for significant concern about the way health literacy might be obtained expressed by secondary stakeholders such as caretakers, physiotherapists or doctors who worried about increased resistance to medical advice in older people. These concerns also referred to inaccurate or false information on the internet and inability to control the quality of information their patients accessed. Our results indicate that doctors worried their professional work would become more complicated with older people who felt more informed. Doctors insisted that it was important for the older people to accept guidance and recommendations and did not see these concerns in conflict with promotion of independence.

For the design of AHA technologies, the challenge lies in addressing such desires and concerns by perhaps creating a space for collaboration and equality with respect to personal health between older people and professionals by means of technology. For instance, health professionals could send personalized health information to their patients' health devices or applications, if requested, instead of older adults relying on information gleaned from the Internet. Technology could mediate the process of improving health literacy in older people, by enabling health professionals to control the quality of health information their patients receive and read.

Reminders as a feature or a problem: Another surprising source of disagreement between primary and secondary stakeholders was reminders – a common feature of all AHA technologies. In our study, many participants were not very fond of this functionality. They understand the intention and the ostensible value in supporting healthy behavior. Yet, receiving constant reminders about what to eat or when to eat, was perceived as the very opposite of independence, being interpreted instead as patronizing. Older adults in our study were sensitive to and wanted control over the frequency and content of reminders. For older adults in our study, reminders were seen as an intrusion, a questioning of their own abilities and at times even an over-emphasis on formal conceptions of what constitutes healthy behavior.

In stark contrast, most secondary stakeholders saw reminders as a useful tool to improve the efficiency of their professional work. They would use such functions frequently to influence the behavior of their patients. Here reminder functionalities were seen as a means to support older adults and to help maintain independence. To the health professionals then, reminders were a way to mitigate their own distrust in the ability of older adults to adhere to medical standards of healthy living. Here independence was interpreted as consistent performance of rote tasks, with reminders used for behavioral modification if necessary.

The design challenge here is to provide a compromise between functionality and user acceptance. While reminders can indeed be useful for older people, we learned that it is important to (1) limit the possibility of professionals to deliver them at any time and (2) take into account the sensitivity of older people towards the frequency and contents of such reminders. For instance, on a very simple level, AHA technologies should provide options to enable, disable or restrict reminders from health professionals in order to provide control to older people. Ideally, reminders could be designed to address individual preferences and personalities of the end users as well (Smith et al., 2016). After all, frequency, presentation and content are highly individual factors and determine whether reminders are perceived as useful or annoying (Bailey et al., 2001; Goldstein et al., 2014; Haberer et al., 2012). These points require collaboration of primary and secondary stakeholders and the mediation and moderation by designers, to find an appropriate ratio of functionality, content personalization and control.

8.5.2 Differing conceptions of well-being

Well-being is an important factor in the success of AHA but we observed that primary and secondary stakeholders partly disagree on what constitutes well-being. For secondary stakeholders well-being is mainly associated with measureable physical and cognitive health, for instance data on physical activity or nutrition.

Here, they tend to follow a deficit approach, focusing more on impairments, limitations and restrictions of older people and how these deficits can be treated with prevention and intervention programs. In contrast, older people in our study associate well-being with enjoying life through means of social participation, independence and self-determination, rather than trying to meet appropriate health measures all the time. There is no doubt, that from a medical point of view nutrition, sleep or physical activity are major factors that influence individual health. Technology supports convenient measurement of these factors through monitoring nutrition intake, sleep patterns and physical activity. These measures then come to be interpreted as evidence of well-being and of behavior that would ensure physical health.

Yet well-being is not merely a result of measured behaviours but an individual feeling and is not compulsory associated with such health factors. Most of current technologies for AHA support follow the same pattern and provide prevention and intervention measures like activity or nutrition monitoring with the stated intention to improve older peoples' well-being. Older adults in our study, however, insisted that well-being was associated with enjoyable activities even if these were clearly less healthy. To them, no amount of healthy nutrition could substitute for the joy of a shared if potentially unhealthy meal. Here we see the distinction between measured and experiential conceptions of well-being. From a design point of view both conceptions ought to be considered thus accounting for the individuality of older people and the requirements of health professionals. By doing so, we might have a greater chance that older people are willing to integrate our technological solutions for long-term AHA support into their daily routines.

8.5.3   Determinants for trust in AHA technologies

Control and trust are important factors for older people, in the context of AHA and technology use (Braun, 2013; Heart and Kalderon, 2013; Lee et al., 2013; Miller and Bell, 2012; Morris and Venkatesh, 2000). Our study shows that older peoples' trust in AHA technologies may strongly depend on the reputation of secondary stakeholders. Seniors are willing to use new AHA technologies under the condition that collected health data is only processed to trusted secondary stakeholders like physicians and that health data processing is transparent. There is a considerable distrust in HIC's, which is induced by the fear of increased health insurance contributions due to the monitoring of unhealthy behavior. In fact, our results show that some secondary stakeholders have indeed a strong desire to exploit health data more largely. However, our data also illustrates that older people imagine the capabilities of HIC's to exploit and use health data to be much more sophisticated as they actually are. Another perspective here is the insight that older peoples' intention to use AHA technologies seems to be heavily affected by recommendations of health professionals. Cimperman et al (2013) also found recommendations of relatives, friends or professionals to have major influence on trust in technologies by older people (Cimperman et al., 2013). For long-term use of technologies for AHA support by older people, it is therefore a noticeable factor that health professionals perceive explicit value by means of improved professional health services and health benefits for older people when using AHA technologies, so they would recommend them to their patients. Having said this, our empirical analysis implies that health professionals tend to refuse AHA technology use by older people for reasons of impaired self-awareness in older people. Such concerns, fears and misunderstandings need to be negotiated among all relevant stakeholders. For instance, AHA technologies should guarantee that under no circumstances personal health data will be distributed to third parties and provide concepts and functionalities to ensure full control over health data by older people. To reduce secondary stakeholders' concerns about impaired self-awareness, AHA technologies should provide a space for the involvement of secondary stakeholders, so that they can keep track of their patients' AHA technology use and communicate measurements to promote self-awareness.

8.5.4   Political will and the policy framework

No matter the quality and ingenuity of AHA technologies, their success and long-term sustainability largely relies on policies and reimbursement. Monetary incentives and financial reimbursements are prominently discussed controversial topics in the area of health promotion (Morris et al., 2004; Volpp et al., 2011). For the success and sustainability of technologies for AHA support in the EU, financial reimbursement seems to be of utmost importance. In most parts of Europe, people pay monthly contributions to the health care system. In return, the health care system pays for necessary health check-ups and treatments. European residents usually expect partly or full compensations for health expenditures. However, in some countries like Germany, Switzerland or France the health care system defines a limited set of health care products and services that are available for monetary compensation. People must pay for products and services, not listed in this catalogue, on their own, which happens rarely for that reason. As designers of AHA technologies, we therefore face two problems: (1) design technical solutions effective and appropriate for the target group and stakeholders (2) ensure sustainable uptake of such technologies within the relevant healthcare system. Due to the structure of healthcare systems in Europe, the latter depends heavily on the willingness of governments to subsidize AHA technologies and provide incentives to people who need such technologies for reasons of prevention or intervention.

Therefore, as designers we have no direct influence on this aspect of the sustainability of such technologies. Instead, the policy maker also paves the way for sustainable use of ICT-based solutions to support AHA in older people.

This is where we come full circle, as sustainable long-term use of AHA technologies demands for the infrastructures and frameworks for successful integration of such technologies into the daily life routines of older people and relevant secondary stakeholders. This includes for instance, financing models with respect to subsidizing AHA technologies for older people, legal frameworks with respect to concepts for data privacy and data control to empower older people, and social infrastructures with respect to participation and integration opportunities, so that loneliness and social isolation become less crucial factors for impaired health. Foundations to such tasks need to be deployed by political stakeholders and policy makers. Designers are only able to take them up by mediating, moderating and negotiating between end-users and all relevant stakeholders in addressing their needs and demands. However, it is the insight of such design activities, presented here that is needed in a first step to (1) derive an appropriate system design and (2) identify recommendations for action, so that in a second step political stakeholders can make decisions and initiate changes that allow for a sustainable impact of AHA technologies on the relevant healthcare system. Therefore, we suggest that the design of AHA technologies for older people should strongly recollect on the strengths and benefits of PD, which provides a space for collaboration and cooperation between relevant stakeholders and enable designers to mediate and moderate between them in order to identify, analyze and negotiate contradictive perspectives, different conceptions and political framework conditions.

8.6   Limitations

The findings described in this article are not generally valid for the entirety of primary end users and secondary stakeholders in the context of the overall healthcare system in Europe. They represent individual perspectives on AHA and technology use. A constant in this work however, is the rationale that careful consideration and integration of all relevant primary end user and secondary stakeholder perspectives is indispensable for the successful design of health-technologies for long-term AHA support. Further, our exploratory interview study did not bring primary end users and secondary stakeholders physically together, which is why we had to take a mediating role. Bringing those stakeholders together might have led to additional implications for the design of AHA technologies. Therefore, our findings illustrate the additional need for a moderating role in PD, to identify and analyze divergent stakeholder perspectives and their interactions to design and develop appropriate long-term solutions for AHA.

8.7 Conclusion

Designers of ICT solutions for AHA support must bring all relevant stakeholders together and collect all different perspectives, to analyze and to negotiate them in order to design and develop meaningful solutions for long-term AHA support. By definition, it is the notion of PD to provide researchers with a framework for collaborative design and development with all relevant stakeholders. This becomes especially relevant with respect to preexisting trust relationships between older people and secondary stakeholders, increased data privacy concerns when using AHA technologies and different conceptions of crucial AHA concepts like independence and well-being. At the same time however, we need the legal framework and infrastructures necessary for sustainable deployment of such technologies into the respective healthcare system. Our exploratory study revealed many contradictions in primary and secondary stakeholders with respect to AHA, which is why researchers in this field need to use PD as a space to mediate and moderate between end-users and all relevant stakeholders in order to build appropriate AHA concepts and technologies for sustainable use.

9 Prototype Design for an Integrative Health Platform to Support Active and Healthy Ageing in Older Adults

Abstract

In this paper, we provide a case study of an integrative health platform for older adults, based on participatory, or co-design, principles. We illustrate the diversity and complexity of older adults' perspectives in the context of health and technology use, the challenges which follow on for the design of appropriate technologies for active and healthy ageing (AHA) support and our approach to addressing these challenges through a participatory design (PD) process. Interviews were conducted with older adults aged 65+ in a two month study with the goal of understanding perspectives on health and technologies for AHA support. We identified challenges and designed a high-fidelity prototype for the integrative health platform "MY-AHA". Finally, we evaluated the prototype with the target group. For researchers in this field, the structured documentation of our procedures and results, as well as the implications derived provide valuable insights for the design of integrative health platforms and AHA technologies in general for older adults.

9.1 Introduction

Life expectancy rates are increasing worldwide as a result of improvements in public health, nutrition, medicine, education and personal hygiene (Active Ageing, 2002; Barry et al., 2015; Centers for Disease Control and Prevention, 2013). There are known problems associated with this, especially in relation to the economic costs associated with an ageing population. Healthcare professionals, for instance, face demands to deliver improved care services with ever more constrained budgets. Thus, there is an increasing interest, among policy makers and health care professionals alike, in approaches and solutions that can help delay the need for healthcare services, prolong independent living, and support older adults in what is now termed active healthy ageing (AHA) (Publications Office of the European Union, 2012; Rechel et al., 2013; Vines et al., 2015). Researchers have responded to the challenge of supporting prevention and the maintainence of healthy lifestyles in older adults through the smart application of technologies (Haluza and Jungwirth, 2015; Stellefson et al., 2015) such as tele-monitoring, remote health services or self-monitoring with wearable devices (Publications Office of the European Union, 2012). Yet there is little evidence that systems oriented towards supporting prevention and health for older adults result in long-term use, which has led in turn to the investigation of methods for improving the sustainability and

© Springer Fachmedien Wiesbaden GmbH, part of Springer Nature 2018
D. D. Vaziri, *Facilitating Daily Life Integration of Technologies for Active and Healthy Aging*, Informationsmanagement in Theorie und Praxis,
https://doi.org/10.1007/978-3-658-22875-0_9

longevity of such systems (Di Pasquale et al., 2013; Jarman, 2014; Wan et al., 2016). Part of the problem is that effective AHA technologies for older adults may require a general understanding of both the varied health needs of older people and the specific circumstances which constrain or afford regular and sustained use. An initial insight is that the health needs and perspectives of older people, while not homogeneous, will nevertheless be distinct in various respects to those of younger people (Buhler et al., 1968; Trujillo et al., 2004). For instance, evidence shows that social aspects like family, social life participation and trust in technologies are strong drivers for AHA technology use in older adults (Ehn, 2008; Fischer et al., 2014; Latulipe et al., 2015). One issue is the need to see frailty in older adults as a multidimensional construct (Puts et al., 2017). Technologies hitherto have nevertheless tended to address limited aspects of overall health by use of e.g. activity monitors, pulse monitors, weight scales, etc. but with little integrated capacity. To support a multidimensional approach for tackling frailty in older adults, we suggest, requires a platform that integrates multiple AHA technologies from different areas of health concerns with the aim of collecting, distributing and processing personal health data according to the various perspectives and needs of older adults.

This design case study aims to investigate how perspectives, attitudes and the real life practices of older adults with respect to AHA can be supported by an integrative health platform. Through a series of interviews and observations we aimed to understand better how older adults use AHA technologies, what prevents and drives their usage of such technologies and how they integrate them into their daily life routines. Based on the results of this participatory design (PD) process, we identified design challenges and addressed them in a high-fidelity prototype for an integrative health platform called MY-AHA. The documentation of our design decisions, procedures, results and prototype evaluation provides, we suggest, implications for the further design and development of integrative health platforms for older adults.

9.2 Related work

9.2.1 Health monitoring and quantify yourself in older adults

The idea of monitoring technologies is relatively new. In 2004, Dishman discussed the potentials of such technologies to support interventions by collecting data on behaviors and detecting problems in a timely manner (Dishman, 2004). Consequently, a vast amount of research in the health domain concentrated on exploiting those potentials by addressing challenges associated with ageing; for instance physical activity, nutritional or cognitive behavior. Today, monitoring technology applied to self-tracking behavior provides the performance and cost-effectiveness to be distributed among a wide user base. Self-tracking devices support the self-management of a variety of life aspects like sleep, nutrition, exercise or mood through the provision of feedback, made possible by recording and analyzing personal health data related to those areas. In general, the provided feedback follows a persuasive strategy with the goal to help users to change their behavior towards a desirable healthy lifestyle (Fogg, 2007). Examples for such devices in the research context are BeWell (Lane et al., 2014), BiFit (Consolvo et al., 2008) and Fish 'n' Steps (Lin et al., 2006). Additionally, commercial applications like Nike+ or FitBit increasingly enter the market. Many of these technologies have been developed for general populations and not older adults in particular. Nonetheless, much research has been conducted in the space of health applications targeted specifically at older adults. Accordingly, developed systems aim to support functional abilities (Lee and Dey, 2011), physical (Doyle et al., 2010a; Uzor and Baillie, 2013), social (Doyle et al., 2010b) or cognitive (Jimison et al., 2010) well-being. Typically, wearable sensors such as pedometers, blood pressure cuffs or pulse oximeters are applied for data collection. However, long-term usage of systems targeting older adults in real environments and older adults' willingness to buy such systems seems limited (Brodaty et al., 2005; Robinson et al., 2009; Wan et al., 2016). Literature suggests that usability and user experience aspects, as well as reliable information channels play a major role in uptake and long-term usage of health-related technologies by older adults (Uzor and Baillie, 2013; Wan et al., 2016). But what seems to be more important when addressing needs of older adults is that technologies for AHA support follow a holistic approach, which includes all relevant aspects of health and well-being, rather than focusing on only one (Thompson et al., 2011). Hence, there is a demand for platforms integrating different devices in order to process health-related data, and orchestrate relevant activities, according to the needs and demands of older adults.

9.2.2 Holistic approaches to support AHA in older adults

Physical impairment is the main hallmark of frailty in older adults (Dent et al.,

2016). However, evidence suggests that other dimensions, such as psychological, cognitive and social factors also contribute to this multidimensional condition (Puts et al., 2017). In current clinical practice, a number of risk assessment tests for early screening of and intervention in physical frailty and cognitive decline are already available, including biomarkers (APOE4, inflammatory markers, Vitamin D measures, etc.), clinical measures (MMSE, GDS, gait, weight, strength, balance, etc.), as well as imaging (MRI, CT, etc). However, while such tests are useful as "single shot" or "threshold" screening tests, few if any are sensitive enough for the identification of early and small changes in levels of risk associated with diseases like dementia, sarcopenia or falls in later life. In addition to measures associated with clinical practice, in recent years many technological solutions for private end-users have been introduced to support early risk detection and to tackle frailty dimensions like cognition (Topolovec-Vranic et al., 2015), physical activity (Gschwind et al., 2015; Vaziri et al., 2017), nutrition (Olson, 2016), sleep (Evenson et al., 2015), emotion (Zhao et al., 2017), social isolation (Khvorostianov et al., 2012) and loneliness (Cotten et al., 2013) among older adults. Despite the fact that AHA technologies for those discrete domains are of a certain value with regard to reducing single risks (fall risk, etc.), there is still a need, we argue, for a more holistic approach which aims to address all of the individual risk factors. Further, and obviously, we need not only to identify a variety of health risks, but also to provide tailored interventions based on the outcomes of risk analysis undertaken across multiple domains of frailty. To realize that, Procter et al. suggest that domain-specific technological solutions like activity monitors or dietary coaches need to be brought together on an integrative platform to analyse health data and orchestrate tailored interventions according to older adults' needs (Procter et al., 2014). However, the diverse requirements of older adults with respect to technology use challenge the design of such platforms in some significant ways.

9.2.3   Design of AHA technologies for older adults

While AHA technologies primarily intend to improve physical and cognitive health, they also affect other aspects of life and *"[...] become the technical infrastructure for a large diversity of different forms of social life"* (Pipek and Wulf, 2009). Therefore, the design of ICT artefacts in the context of healthcare innovation, we argue, needs to consider and understand the social arrangements of older adults who apply these artefacts. Gaining such understandings, moreover, necessitates approaches which place designers and users in equal positions (Rohde et al., 2009). However, achieving equality in the design process turns out to be a challenging process. The literature mentions several challenges in the design of

AHA technologies for older adults, for instance, (1) older adults' limited capability, in some cases, to understand technical terms or artefacts and articulate requirements and obligations, (2) their longer learning curve and need for multiple iterations to get used to AHA technologies and (3) their need for social support infrastructures not only for technical issues but for social participation (Chaudhry et al., 2016; Eisma et al., 2003; Lindsay et al., 2012). To address such challenges, research has indicated the need for methodologies that support a better understanding of technological requirements for the design of technologies appropriate for older adults (Keith and Whitney, 1998; Lindsay et al., 2012). Traditional technology design approaches, therefore, need to be complemented with a thorough investigation into users' needs and their everyday lives and practices in order to develop sustainable, usable and useful technology for this target group (Ebert and Heimermann, 2004). This calls for an active collaboration in the design process between older adults, researchers and industry (Newell et al., 2006). Participatory Design (PD) constitutes an appropriate framework to identify and address design challenges in work with older adults and provides for a methodological eclecticism to support a better understanding of older adults' specific needs. In the context of AHA technologies, PD has been successfully applied in various studies (Grönvall and Kyng, 2013; Lindsay et al., 2012; Siek et al., 2010). Even so, and depending on what kind of participation is envisaged, PD can be a long and complex process and providing access to structured results and procedures of AHA technology design for other researchers and developers can be challenging.

9.2.4   Design case studies in AHA technologies
Once one has accepted the need for a participatory approach then it is obvious that gaining an understanding of the social practices of older adults and making sense of the possible scope of proposed designs and, ultimately, specific design choices, needs to be coordinated to develop ICT healthcare innovations that integrate seamlessly into the real world contexts of older adults (Wulf et al., 2011). In order to clarify the complex choices designers have to make and to explain their thinking behind user experience and visual decisions, the use of design case studies may be an appropriate approach here (Papanek, 1983; Wulf et al., 2011). Based on the rationale of case studies, Wulf et al. (2011) developed a research approach composed of three phases, (1) empirically analysing given practices in a specific field of application, (2) designing an ICT artefact, based on the findings in phase 1, (3) examining the appropriation of the ICT artefact with the target group over a longer period (Wulf et al., 2011). They called the documentation of such a three-phase research approach a *design case study* (Wulf, 2009). Existing research into iStoppFalls, a European funded project to design an ICT-based fall prevention system for older adults, serves as a good example for a design case study in the context

of ICT healthcare innovations (Ogonowski et al., 2014; Vaziri et al., 2016, 2017). The results illustrated the detailed insights and in-depth implications design case studies may provide for the design of AHA technologies. Still, iStoppFalls mainly provided exergame-based solutions for fall prevention and therefore was only able to address demands and needs of older adults interested in such interventions. As a partial solution to frailty, iStoppFalls could not account for the heterogeneity of older adults. The solution presented in this paper proposes a platform that integrates partial solutions like iStoppFalls and adjusts them to the heterogeneous demands and needs of older adults.

9.2.5   Research question
Relatively little research currently addresses the challenge to design a holistic and integrative ICT-platform to support early risk detection and AHA in older adults across relevant areas of concern. Many currently available technologies for AHA support address heterogeneous needs and demands of older adults insufficiently and therefore older adults use these technologies short-term only, which is one reason why anticipated health improvements stay out (Fitzpatrick and Ellingsen, 2013; Wan et al., 2016). In order to understand why that might be, we argue that there is a pressing need to undertake a thorough investigation of attitudes and practices of that target group with respect to health and technology use. However, research in the ICT healthcare context investigating daily life practices and perspectives of older adults for AHA technology design mainly do so in the light of technological solutions that only attempt to tackle frailty in older adults partially, for instance from a physical or cognitive perspective. With our research we address three questions relevant for the design of integrative health platforms; (1) how do older adults perceive health and relevant technologies in their daily lives, (2) how may the heterogeneity of older adults be addressed by designers in early design stages and (3) what implications may be valuable for further design of AHA technologies for older adults.

9.3   Methods

9.3.1   Research setting and study setup
The study is part of the European research project MY-AHA, the goal of which is to design an ICT-based integrative health platform for primary end users and secondary stakeholders. The MY-AHA platform will be designed as an Android App for smartphones and tablets. It will allow the users to install and connect a range of health software applications and AHA technology devices like activity monitors, dietary coaches, etc. to it, in order to tackle frailty in six areas of concern, (1) physical activity, (2) cognition, (3) emotion, (4) nutrition, (5) sleep and (6) social

life. The platform will process health data collected from installed applications and connected devices. Based on that data, the users will be provided with calculations of individual health risks and tailored prevention or intervention programs that address the needs and demands of the target group. Further key aspects of MY-AHA will entail the sharing of health data with other stakeholders (for instance doctors) and connecting users with each other to participate in programs together. According to the design case study methodology (Wulf et al., 2011), we organized our study into three steps.

Firstly, we conducted a pre-study over a period of two months in order to understand the perspectives, attitudes and practices of end users with respect to health, illness and injury prevention and technology use. Therefore, we initially conducted a workshop with all participants to explain the goal of the European research project, the purpose of the underlying study and technology-based self-assessment.

Fig 13. Initial workshop with pre-study participants

Afterwards, we introduced to our participants various self-assessment technologies like activity, pulse and sleep monitors, pulse oximeters, weight scales, blood pressure cuffs and glucose meters, provided by Medisana, a commercial company producing AHA technologies. We explained the parameters measured by these devices, elaborated on potential benefits when using them and illustrated their handling. Subsequently, participants were requested to try out the introduced devices. Thereafter, we put up different stations for each device, equipped with a smartphone and tablet to enable access to visualizations, (for instance, results). A trained research assistant supervised each station to encourage participants to try out the devices and to answer questions arising. At the end of the workshop, participants were requested to choose the devices they would like to use over the next two months. They were allowed to choose multiple devices.

Fig 14. Introducing and explaining various technologies for AHA support

A second step focused on the design of a high-fidelity prototype, based on pre-study outcomes and through examination of the available literature in this field. The prototype, we felt, should reflect operative functionalities and the perspectives, attitudes and practices of older adults collected during the pre-study phase. The design process followed three stages. The first stage described primal control elements with wireframes and their interplay with respect to logical interaction. In a second stage, mock-ups were developed which illustrated the visualized appearance of the application, including for instance a color scheme, typography, and icons. For visual appearance, we followed the design framework from Google's material design, mainly to provide a consistent user experience on mobile devices. In addition to Google's material design concept, we addressed accessibility and understandability aspects, as well as the practices of our target group identified during the pre-study. To generate the graphical mock-ups, we used the design and prototyping tool *Sketch*. Finally, in a third stage, we translated the graphical mock-ups into an interactive prototype to simulate a real world application. The focus of the interactive prototype was simulating a coherent application in which the user is able to navigate. Therefore, we used the prototyping tool, *Axure6*, which specializes in the conception of dynamic elements and processes.

In the final step of our design case study, we conducted a prototype evaluation of the interactive prototype with participants, moderated by a trained research assistant. We defined two scenarios with a duration of 90 minutes each

(Barnum, 2011). Scenario one included six tasks for the tablet version, while scenario two included six tasks for the smartphone version of the interactive prototype. The tasks for the tablet version requested participants for instance, to navigate the application freely, to look up the connected devices and install third party applications, to check and adapt their daily steps goal, or to look up their monthly activity statistics. For the smartphone version, we asked participants, for instance, to navigate to the home screen and describe what they see there, to check for their risk parameters and collect information on how to improve them, to look up their currently registered interventions to lower physical risks, or to deactivate health notifications. Smartphone and tablet tasks differed to cover a wider spectrum of prototype functionalities and to observe interactions with both devices, unbiased by experiences made with the previously used device.

9.3.2   Participants
Initially, the study included seven older adults (four male and three female) aged between 65 and 85 years. These participants lived independently at home and participated in our two month pre-study. For the evaluation phase, we recruited five additional participants (two male and three female) aged between 65 and 86 years to increase the robustness of our evaluation results. While four of the new participants were residents of a residential home, one participant lived independently at home. For the total sample of twelve participants, one-half was in a healthy physical condition with respect to their age, meaning no signs of frailty. The other half showed symptoms of frailty, for instance arthrosis or impaired balance. All participants were recruited from two cities in Germany; Cologne and Siegen. With respect to their experiences with mobile technologies before the study, our sample provided a wide spectrum, ranging from older adults without experience of mobile technology to older adults with advanced technology experience. Three older adults already had experiences with AHA technologies prior to the study. Table 9 provides a description of all participants. Seven Participants (PN 1-6 and PN 8) participated in the pre-study. After the pre-study PN 2 and 3 dropped out for reasons of insufficient time. Therefore, ten participants (PN 1 and PN 4-12) took part in the evaluation study of the interactive prototype.

Tab 9. Description of participants

No.	Gender	Age	Health status	Mobile technology experience	AHA technology experience	Pre study	Evaluation study
PN 1	male	74	physically impaired	advanced	none	yes	yes
PN 2	female	71	physically fit	medium	none	yes	no
PN 3	male	71	physically fit	medium	pedometer	yes	no
PN 4	female	74	physically fit	low	none	yes	yes
PN 5	male	78	physically fit	low	none	yes	yes
PN 6	female	85	physically impaired	none	none	yes	yes
PN 7	male	68	physically fit	advanced	sleep monitor	no	yes
PN 8	male	65	physically fit	medium	pulse monitor	yes	yes
PN 9	female	86	physically impaired	none	none	no	yes
PN 10	female	86	physically impaired	none	none	no	yes
PN 11	male	65	physically impaired	medium	none	no	yes
PN 12	female	86	physically impaired	none	none	no	yes

9.3.3   Data collection and analysis

Pre-Study: At the beginning of the pre-study, participants were given a booklet to document their experiences with the devices over the following two months. We requested them to document their experiences once a week with respect to frequency of usage, functionalities used, integration into daily life routines, technical problems, perception of individual customizability, configuration options utilized and desired configuration options to adapt the device to their needs. Further, at the end of our two-month pre-study, we conducted semi-structured interviews with participants with a view to assessing their experience. For instance, we asked them (1) for their general experiences regarding the used devices made during the last two months, (2) for their willingness to use such systems in future, (3) whether using these devices affected their health-related behaviour, (4) if they have specific desires or recommendations for the development of MY-AHA, (5) how they decide whether to use a health device or not, (6) how they perceived the possibility of setting goals, if goals had a motivating character and how they used them, (7) if they required support to use the devices, (8) how they perceived notifications and reminders and (9) their attitude towards data privacy and sharing of personal health data. Participants were allowed and required to elaborate freely on those topics. Two trained research assistants conducted and moderated all interviews. Each interview was audio-recorded and afterwards transcribed.

The data material was analysed by applying a thematic analysis approach (Braun and Clarke, 2006). Based on the transcribed audio files and booklet documentations four coders performed an inductive analysis of the data material and generated main categories. Coding discrepancies were discussed and eliminated by adding, editing or deleting codes, based on the group discussion outcomes of the coders. The final code system covered categories relating to participants perspectives on health and prevention and how these can be supported by self-assessment technologies, their perceived benefits and drawbacks of self-assessment technologies, usability aspects, the perceived role of self-assessment technologies, effects on personal well-being induced by self-assessment technologies and trust aspects affecting the use of these technologies. Further, we developed relevant personas, which condensed the substance of the interviews and highlighted different user perspectives with respect to health, prevention and the use of self-assessment technologies. Based on the material analysed, we derived implications for the design of the prototype. Coders used the software application MAXQDA for the thematic analysis.

Prototype evaluation: For the prototype evaluation, we conducted usability tests based on the twelve tasks described in the section above. We provided a document including these tasks for each participant. Additionally, the moderator was given the same document, which further contained questions related to specific tasks. For instance, the moderator asked participants who were completing the respective task, for their comprehension of terms like "App" or "risk factors", what they expected when they might hit the share button, how they perceived specific elements of the interface, what information was missing in their opinion, and if they would be willing to provide the data necessary to sign up to the application. In addition, all participants were requested to think aloud while completing the tasks. Each of the ten evaluations was conducted successively. Subsequent to the prototype evaluation, we conducted a final interview with each participant, in which we asked about possible usability or accessibility problems, if they believed they could easily learn to use the application on their own, whether or not the system provided an appropriate degree of support with respect to individual capabilities of the user, if the application provided additional value to the user and if they could imagine using such a system regularly. To conclude the interview, we asked participants for any remaining aspects or recommendations they would like to provide with respect to further development of the system.

The analysis was conducted in the same way as the pre-study analysis. Categories related to the application structure, visualization of health data and results, perception of self-assessment, perspectives on data privacy and misuse of data, integration into daily life routines, used terminology, and ease of use were derived.

9.4   Pre-Study Findings

9.4.1   Relevant themes for AHA technology use by older adults

Our pre-study provided detailed insights into relevant themes for AHA technology use by older adults. In total, six themes were identified, (1) perspectives on health and prevention, (2) context of usage, (3) benefits and drawbacks of AHA technology use, (4) Trust in self-assessment technologies, (5) the perceived role of self-assessment technologies and (6) usability and accessibility aspects.

1. Perspectives on health and prevention: From our pre-study, we learned that older adults understand health and prevention primarily in light of physical activity and social participation. All participants understand the need for prevention as a means to maintain social connectedness and most welcomed the idea of technology to support this prevention activity: *"Well, I need social contacts! [...] This is why I want to stay in my own home as long as possible, and of course, I appreciate any technology that can help me with staying healthy longer [...]." (PN 6, female, 85 years).* Further, some participants saw a distinct association between social participation and physical activity that would drive them to be more active. *"[...] I like to go to the gym. But, when you go alone all the time you easily find excuses not to go. And such excuses will become more and more common. That is why I prefer to go to the gym with others, also to compete and put more effort. It is a psychological effect." (PN 6, female, 85 years).* In addition, participants admitted that social support would increase their motivation for healthy behavior: *"yes of course. They [family] would see it [increased physical activity behavior]. I know that this would motivate me to keep on going." (PN 7, male, 68 years).*

2. Context of usage: For most participants, usage context played an important role for the choice of self-assessment devices. In general, they chose technologies that would support them in existing prevention activities or current health problems, for instance weight monitoring or running activities. *"I already adapted my diet before. This [self-assessment devices] is just a control mechanism. The devices just came at a proper time." (PN 8, male, 65 years).* By using the devices, some participants raised interests in new health topics that they had not considered before. In answer to a question concerning what she anticipated by continuing to use the device, a female participant responded: *"[...] that I try to find out more about the causes. For instance, why did it take me so long to fall asleep? Why did I wake up several times? I was able to find out more about some of these topics, thanks to the measurements and information provided by the device [sleep monitor]. Hopefully, I will learn more by continuing to use the device." (PN 6, female,*

85 years). In contrast, a male participant complained about insufficient information and lack of assistance from the sleep monitor device. *"well, I don't know which ones [interventions] to initiate. [...]. The device did not make any recommendations here, to tell me what I could improve." (PN 2, female, 71 years).*

3. Benefits and drawbacks of AHA technology use: The use of the self-assessment technologies over a period of two months noticeably affected participants' health behavior. Positive effects described by the participants ranged from increased health literacy and security to increased motivation and discipline with respect to healthy behavior. One participant stated: *"[The devices] discipline one to be more active. They motivate you to do more and keep on going." (PN 8, male, 65 years).* PN 4 mentioned that using the devices brings her pleasure. *"I would enjoy it, if I could continue monitoring that [her daily step goal]." (PN 4, female, 74 years).* Besides positive effects, some participants perceived the use of self-assessment technologies in a more negative light. One participant expressed their concerns about unwanted observation. *"Well, I like those devices most that are easy to operate and don't give me the feeling of being observed 20 hours a day:" (PN 3, male, 71 years).* PN 5 was concerned about the effort required to use such devices properly and doubted their overall value. *"I have to compare the required effort with the benefit and I see a certain discrepancy here. [...]. Also, the reasonability of assessing my behavior with such devices is very debatable." (PN 5, male, 78 years).* PN 1 warned about the power of measured health data and its possible consequences. *"[...] I am very concerned that my attention is focused too much on these things [measured parameters]. For instance, when I measure my pulse and I don't understand the parameter correctly, I become insecure." (PN 1, male, 74 years).*

4. Trust in self-assessment technologies: Throughout our interviews, most participants emphasized the importance of trust in self-assessment technologies, where they defined trust as a combination of reliability (measurement accuracy) and privacy (data security) afforded by health devices. Over the course of the study, the users had different reported experiences with respect to reliability. For instance, PN 8 was positively surprised by the device's accuracy. *"The measurement is quite accurate! I did the annual health check in the same period and my doctors' results matched the results from the device." (PN 8, male, 65 years).* According to our observations and interviews, that experience positively affected PN 8's willingness to use the device and increased their usage intensity. In contrast, some participants reported negative experiences with respect to reliability. *"They were wrong. I definitely did not take so many steps. The devices tracked my steps, even when I was sitting on the couch. You cannot honestly tell me that this is right." (PN 2, female, 71 years).* For that reason, PN 2 contested the meaningfulness and usefulness of such technologies and refused to continue using the pedometer for the rest of the study. In general, our observation and interview data indicated- perhaps unsurprisingly- that positive reliability experiences intensified the use of AHA technologies, while negative reliability experiences increased frustration and led to decreased usage intensity or even usage abortion.

In terms of privacy, more than half of our pre-study participants emphasized the significance of data security in general and in the context of health-related data collection. While this is also not a surprising result, the reservations of the target group with respect to AHA technology use revealed specific functionalities they would or would not use respectively. Here a major concern is the transfer of health-related data, for instance to a doctor, directly via the health device. *"No, I would insist on analyzing the data by myself first, and then deciding which data to transmit to my doctor. I want to be the owner of my data, [...]. I would never transmit my health data through an external platform or a health device." (PN 1, male, 74 years)*. In this context, PN 1, 3 and 5 explicitly mentioned that they would prefer a non-digital means to share their data, for instance by printouts.

5. The perceived role of self-assessment technologies: From the interviews, we learned that the interaction modes between the system and the user, as well as the perception of the system by the user are important to older adults when deciding whether to use self-assessment technologies or not. Accordingly, some participants were quite skeptical with respect to support given and proactive assessments by the system in technology-based prevention programs. *"[...] statistically you say a body-mass-index of 24 is good. I don't like this development. Being dependent on some technology-based measures [...]. I feel pressurized. I rather listen to my body and assess my well-being by myself." (PN 4, female, 74 years)*. PN 1, PN 6 and PN 8 share the concerns expressed by PN 4. They further mentioned that they disliked being patronized by the system, for instance by receiving reminders frequently, even though they otherwise felt healthy. However, PN 4 also stated that such supportive functionalities would be helpful in other areas where he had less experience, for instance nutrition. The remaining participants were less skeptical and perceived reminders and notifications as helpful. PN 2 observed that such a system-support provided comfort, as it prevented misperceptions of oneself. *"This [notification if set goals are appropriate] would definitely be a reasonable function. If the system tells me that my set goal is too ambitious, I would be reassured not to overstrain myself." (PN 2, female, 71 years)*. Overall, it seems that support and proactive behavior are sensitive and complex topics in the perception of older adults, which need to be addressed accurately and appropriately in relation to the demands of that target group.

6. Usability and accessibility aspects: The assessment of usability and accessibility aspects is strongly related to a specific technology and has been investigated in many other studies before. Therefore, usability and accessibility problems identified by using the Medisana devices cannot be completely detached from the tested devices and only provide a limited contribution to the design of an integrative health platform such as MY-AHA. However, these specific device-related issues allowed us to derive more general aspects that seem to be relevant for the overall use of AHA technologies by older adults. In this context, the integration of self-assessment tools into daily life routines was a major aspect for our participants. For them the integration was impaired by insufficient usability and accessibility. *"I could imagine using such a device [Pedometer] on a daily basis. However, as I said before, the device I tested was very cumbersome to adjust and operate. This takes too much time for me. [...]. Also examining the results is not very intuitive."* (PN 1, male, 74 years). PN 4 complained that results lacked explanatory descriptions and required the user to find missing information through the internet. *"Yes yes. Of course, I could get the missing information about calories on the internet, but this would be very inconvenient to me. I want the information at hand when using the device."* (PN 4, female, 74 years). Finally, participants who had no prior experiences with AHA technologies expressed their desire for detailed and easy to understand manuals and instructions to reduce learning time and prevent potential frustration. *"I don't remember where to set the target for the day. I need instructions how to get to that particular screen. As I am afraid to push any buttons and delete all data, I rather stop using the device (PN 4, female, 74 years).*

9.4.2 Identified personas

The previously mentioned themes illustrate the complexity of AHA technology use in older adults. That complexity is defined by the interaction of older adults' perspectives on health and trust, technology-related aspects such as usability or usefulness and the perceived role of a system that needs to adjust support and communication functionalities to the demands and capabilities of users. In this context, we need to be reminded that attitudes, practices and behaviors of older adults are highly diverse. To reduce complexity, we developed personas, based on essential and marked differences between our participants. These personas highlight participants' viewpoints crucial for the design of an integrative and configurable health platform. In total, we derived five personas based on pre-study interview material, (1) the performance-minded user, (2) the worried user, (3) the needy user, (4) the skeptical user and (5) the interested user.

1. The performance-minded user: This user considers prevention and self-assessment to be largely a function of physical activity. Based on a high degree of self-initiative and competence, the user applies technology primarily for the performance measurements and quantification of specific health parameters. Goals may incentivize the frequent usage of the system. Physical activity behavior comes from an intrinsic motivation and minor technical problems do not impair the user's willingness to use the system regularly.

2. The worried user: This user tends to worry a lot about their own health. Concern over age-related indicators and the increased prevalence of certain diseases means that the interest in prevention programs is substantial. For the good of their own health, this user is ready to deal intensively with health-related measurements. For this user, it is crucial that technology provides self-explanatory health information and thereby induces a palliative effect on the user. Maintaining self-control and autonomy are important aspects for this user in initiating and continuing the use of AHA technologies.

3. The needy user: This user is characterized by passiveness and low self-initiative with respect to their own health. Therefore, expectations towards technology-based support for preventive measures in daily life are especially high. This concerns the need for assistance for technical, training and motivational aspects. Triggers for extrinsic motivation and a perceivable benefit of technology use in daily life are necessary to maintain technology use by this user. Minor technical problems or incomprehensible measurements may cause frustration and refusal.

4. The skeptical user: Data security and privacy are top priorities for this user, which is why they are very skeptical about technology-based prevention and self-assessment systems. Willingness to share personal data with such systems is extremely limited. Further, this user challenges the meaningfulness and reliability of preventive measures and self-assessments. Required efforts to apply such systems and the request to share personal data are distinctly evaluated against potential benefits for personal well-being. A loss of trust will inevitably lead to the termination of AHA technology use.

5. The interested user: This user considers technology-based prevention and self-assessment a chance to deal with information and communication technology to improve their own competence. Conducted by the goal to follow technological trends in old age, this user has a large learning receptivity to appropriate ubiquitous devices. They require either social or technical support, to compensate their low technology competence. Here, simplicity of technology and a step-by-step approach are key.

In the context of the following prototype evaluation, the personas will help us to discuss findings in regard to older adults' heterogeneity and understand how our design artefact addressed their needs and requirements.

9.4.3 Design challenges

The reported pre-study results provided detailed information on how older adults perceive technology in the context of their own health. In the light of the heterogeneity of older adults' practices and attitudes, four design challenges for MY-AHA emerged, (1) usability, (2) intelligibility, (3) usefulness, and (4) sustainability.

1. Usability: Usability aspects should be prioritized in the design of graphical user interfaces and processes in order to ensure predictable system behavior, as well as adequate display, font sizes, and color contrasts. Moreover, the design of the platform should facilitate reliability and a sense of compatibility by aligning it to the design of known and trusted applications of older adults, for instance google or android applications.

2. Intelligibility: Low levels of experience with mobile and ubiquitous technologies in our participants led to the demand for didactic usage support. This comprises the possibility to learn use of the system through a step-by-step approach or of making use of explicit manuals when necessary. In this context, participants often failed to find essential menus like system options and therefore wished for support to locate such menus more easily. Further, they expressed an appreciation of the availability of a personal contact, to help them with technical problems.

3. Usefulness: Individuality in the context of prevention programs is essential for the perceived usefulness of the system. Participants associated individuality with the desire to act autonomously and in accordance with their own competencies. Therefore, users normally want to choose prevention programs rather than have certain assistive or monitoring technologies mandated. Technology, they felt overall, should only provide information on the purpose or the operation of prevention programs, applications and devices. Further, participants' perception of a technology's usefulness seemed to be influenced by the notifications, suggestions or feedback provided by the system. While some participants welcomed proactive support such as interpretation of measurements, or their comparison with benchmarks, praise and admonition or incentives, other participants felt patronized and stopped using the AHA technologies. Such perceptions can be ameliorated by making the MY-AHA platform configurable in such a way that users can easily determine what advice, support, encouragement or otherwise they want and need. Further, proactive support by MY-AHA needs to be sensitive to the impact of wordings, especially in the context of negative messages.

4. Sustainability: Seamless integration into daily life activities of older adults is key for the participants if they are to use AHA technologies long-term. Here, three major aspects were the compatibility of MY-AHA with other mobile technologies already in use, for instance smartphones or tablets, the possibility of providing controlled health data access to trusted doctors for the purpose of diagnostics and the unobtrusive and easy use of MY-AHA to reduce daily effort. Additionally, trust seemed to be an important factor for sustainable use of AHA technologies. Participants mentioned their reluctance to begin or continue using AHA technologies if too much information is requested or data processing is non-transparent. Therefore, in order to promote trust in the MY-AHA platform, information required to sign up to the platform should be kept at a minimum and transparency in data transfer processing and full control over data transfer procedures have to be provided to older adults.

9.5   MY-AHA platform prototype

We addressed the design challenges in a first prototype version of MY-AHA. As our pre-study and literature suggests, applications for older adults need to have a low degree of complexity that supports localization of contents and simple navigation within the application and is further seen as both reliable and useful. In order to reduce complexity for our participants, we decided to use the content design presented in figure 15 for both devices. However, due to the different display sizes, menu visualization on the smartphone version differs from the tablet version (see figure 15).

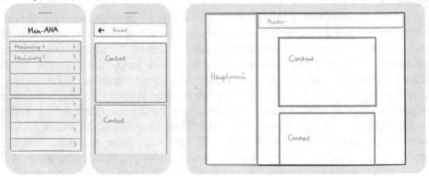

Fig 15. Smartphone and tablet menu

Key menus presented within the MY-AHA platform are the dashboard, applications, devices, results, goals, risks and interventions. The dashboard provides a summary of the latest results regarding activities performed by the user. The applications menu lists third party applications available for instalment or already installed on MY-AHA. Likewise, the device's menu lists third party devices, for instance activity or sleep monitors available for connection, or already connected to MY-AHA. The results menu provides information on results achieved from different areas like physical activity, nutrition or cognition. The goal settings menu allows the user to set personal goals such as loss of weight, a specific number of steps, distance to be covered over the day or daily food intake. Within the risk menu, users will find their individual risk for diseases or health hazards, based on their personal health data and behaviour. Finally, the intervention menu lists currently active interventions, for instance fall prevention or dietary plans, and provides suggestions for additional interventions, based on the user's health condition. In addition, the platform provides menus for profile settings, messages and system options. In the profile settings, the user can adjust personal data like age, height, blood pressure or educational years, which will flow into the calculation of risk models among others. In the messages menu, the user can check personal messages from friends, family or other platform users. Finally, the system options menu allows the user to adjust settings for notifications, reminders, feedback or updates.

To organize the required menus within the MY-AHA application, we had to choose between two different menu structures, a flat menu structure with only one level and a nested menu structure with multiple levels (see figure 16). Literature suggests that flat menus may better address the characteristics and limitations of older adults, for instance cognitive limitations or problems with navigation, which is why we made a design decision for the flat menu structure (Carmien and Manzanares, 2014; Gudur et al., 2013; Lorenz and Oppermann, 2009; Torun et al., 2012).

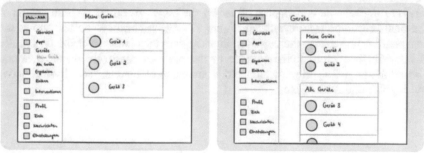

Fig 16. Nesting versus flat menu structure

Subsequent to completing the wireframes for all menus, we transformed them into visual design artefacts. Key aspects in the visual design focused on color scheme, control elements, typography, icon design and presentation of health data. Material design suggests selecting one primary color for the header and control elements of an application (Google, 2017a).

Studies have shown that blue seems to be the most universally accepted color and is associated with trust, loyalty, integrity and honesty (Mendoza, 2009;

Minnick, 2016). Such features are crucial for our target group, which is why we chose blue to be the primary color for MY-AHA.

For control elements, material design suggests the use of flat or raised buttons (Google, 2017b). The main difference here is the affordance. We refer to the term affordance as a meaningful indication about what actions are possible, and what they should do next according to their own action capabilities (Awad et al., 2014; Michaels and Carello, 1981). Therefore, an application design appropriate for use by older adults should aim for strong affordances. Compared to raised buttons, flat buttons provide a weak affordance, which is why we selected the former for our visual design.

For the applied typography concept, we followed material design's recommendation, to use *roboto* as standard typeface on Android devices. However, to address specific capabilities of our target group with respect to impaired vision, font sizes for heading one and two, as well as for body text were increased by 3pt each.

Regarding the icon design in MY-AHA, studies on learning new technologies provided evidence that the use of icons with text labels or text labels alone supports initial learning by the user, while the use of icons alone impairs the early stages of learning (Wiedenbeck, 1999). Further, the use of icons with text labels showed positive correlations to perceived ease of use in new technologies (Wiedenbeck, 1999). Therefore, MY-AHA uses a combination of icons and text labels to improve perceived ease of use by the target group. The visual design of the final prototype is illustrated in figure 17. The picture on the left shows the dashboard on a tablet, summarizing physical activity results over the past seven days, while the picture on the right shows the entry screen for the results menu on a smartphone, listing all relevant health domains in MY-AHA like physical activity, cognition, nutrition, etc.

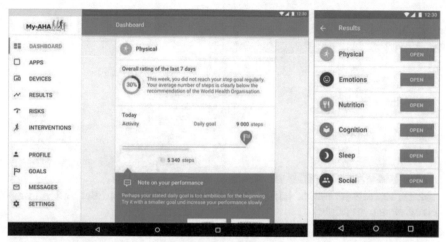

Fig 17. Dashboard on tablet (left) and entry screen for results menu on smartphone (right)

Further, a crucial aspect in the design of MY-AHA was the visualization of health data with respect to achieved results and health risks, as well as proactive messages provided by the system. Regarding health data presentation, the target group wished for a simple visualization of results with absolute measures only, and easy to understand indicators such as traffic light schemes. Regarding proactive system messages, participants in our pre-study emphasized that being patronized, for instance through inappropriate wording or annoying application of such messages should be avoided. The left picture in figure 17 illustrates an exemplary notification with respect to the user's performance, indicating that the set goal might be too ambitious. The user may choose whether to adjust the goal (right button) or hide the notification (left button). MY-AHA provides four different types of messages, (1) health related suggestions, (2) motivational feedback, (3) reminders and (4) tips for system use.

Figure 18 shows our attempt to implement requirements regarding results visualization in MY-AHA. The left picture visualizes a result screen on a tablet, summarizing covered distance in steps over the past seven days with a traffic light scheme, indicating poor, mediocre and good performance. Additionally, the recommended level of steps by the WHO is displayed on the button of the screen. The picture in the middle shows a result screen on a smartphone, outlining covered distance in steps over the past month. The connected dots on the left side represent the actual performance, while the dashed line on the right with a flag on top represents the set goal. The picture on the right illustrates two alternatives for summarizing results on an abstract level.

Fig 18. Results screen on tablet (left) and smartphone (middle) and summarized measures (right)

Finally, individual health risks represent a complex calculation, based on health data collected with different devices such as activity monitors, sleep monitors or dietary applications, and therefore need to be visualized in a comprehensible way. We designed two alternatives for health risk visualization (see fig. 19). The picture on the left shows a coordinate system with age on the x-axis and risk value on the y-axis. Thresholds, derived from scientific literature, indicate the individual risk from very marked to very low. The picture on the right illustrates a more abstract view on health risks, by simply categorizing the individual risk into three segments (green, yellow and red) and indicating the current risk with a needle and in form of an absolute number (1.5).

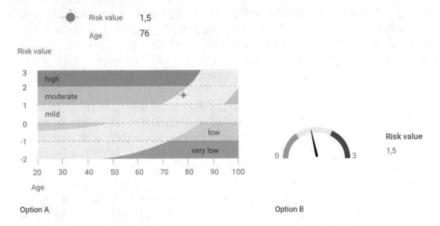

Fig 19. Health risk corridor (left) and health risk tachometer (right)

9.6   Prototype evaluation results

As illustrated in the previous section, we attempted to address the design challenges identified within the pre-study in a prototype version of MY-AHA. In order to evaluate whether design challenges were addressed appropriately and to uncover remaining problems, we conducted usability tests with our participants. We deal with the results of these tests in what follows.

9.6.1   Usability

From the interview and observation material, we identified five different usability aspects, (1) structure of the application, (2) interactive elements and their appearance, (3) icon design, (4) color scheme and typography concept, and (5) visualization of health data.

1. Structure of the application: In general, participants, especially the younger participants in our cohort, did not encounter any substantial problems with respect to navigation and orientation in both tablet and smartphone versions. In this context, PN 8 stated that *"[The application] is clear and easy to understand. I understand the intention of the application very well". (PN 8, male, 65 years)*. However, the tests showed that older adults aged 80+ years struggled to navigate between specific screens. According to our observations, main reasons were a strong focus on the content area, overextension caused by the complexity of the application and reluctant click behavior. For instance, we observed that PN 6 only kept the current screen in focus, so that the moderator needed to remind the participant that there are other screens and menus available. While navigating through a different screen, as PN 6 remarked, it is difficult to bear in mind the part of the application that is not visible on screen. With respect to the allocation of functionalities into the different application menus, participants did not experience noticeable problems, with exception of the goal-setting menu. While, PN 6 and PN 9 did not find the goal-setting menu, PN 5 was able to find the menu right away. All remaining participants searched several other screens such as the results, interventions or devices screen first, indicating that goal-setting was not expected to be found in a separated menu.

2. Interactive elements and their appearance: The tests showed that all participants perceived the raised buttons as interactive elements. However, on the entry screen for the results menu (see fig. 5, right picture), five participants first clicked the domain name and the intended button afterwards. With respect to other interactive elements, such as sliders, check boxes or radio buttons, mainly used to visualize health risk simulations, some participants did not perceive these elements as interactive. Further, the meaning of the switch element, used to turn on or turn off proactive notifications and suggestions, was inconclusive for PN 1 and PN 6. While PN 1 thought using the switch would deactivate the whole domain, PN 6 assumed using the switch would turn off the device.

3. Icon design: As expected, the combination of icons and text labels, as illustrated in figure 17, was appropriate for and appreciated by the target group. Participants were able to distinguish clearly between domains, wherever icons and text labels were shown together. However, due to space problems, especially in the smartphone version, we decided to use icons without text labels for the applications, devices and interventions screens. Our tests showed that more than half of our participants (PN 1, PN 4, PN 5, PN 6, PN 7 and PN 10) were not able to recall the meaning of the icons and therefore interpreted them wrongly or misperceived them as interactive elements.

4. Color scheme and typography concept: With respect to the color scheme used, participants provided positive feedback. In general, our observations and interviews showed that the color-coding, used to distinguish contents within the application, supports the visual assumptions of our target group. In addition to the color scheme, the textual design of proactive notifications and suggestions attracted the participants' attention in a positive way. In this context, PN 4 said, *„I enjoyed reading the notifications and suggestions. They were phrased in a very friendly way. That is important!" (PN 4, female, 74 years).*

5. Visualization of health data: We provided different alternatives for visualizing health data and risks in MY-AHA (see fig. 6 and 7). With respect to the period deemed suitable for results (see fig. 6), participants preferred a weekly timeframe to a monthly, as it seems to correspond to the target groups' behavior. In this regard PN 7 stated, *"One week is ok, it really is. It is appropriate, as you live in such intervals". (PN 7, male, 68 years).* For the visualization of these results, participants praised the selected color scheme. PN 10 said, *"It [results] catches the eye, as it is bolder and I can see at first glance, if I performed well or not". (PN 10, female, 86 years).* In the same context, PN 4 stated, *"I like it [color scheme]. When I performed very well, it shows a different color. I am always surprised when that happens. Using colors [to visualize results] is a good idea". (PN 4, female, 74 years).* To summarize health data on an abstract level, participants preferred either a circle or a combination of circle and smiley (see fig. 6), as they perceived percentage values as more informative, accurate and meaningful. None of the participants would prefer the exclusive use of a smiley. To visualize the individual health risk, the tachometer is preferred over the coordinate system (see fig. 7) for all participants. In fact, four participants misinterpreted the coordinate system, while only one participant misinterpreted the tachometer. In this context, PN 1 said, *"Even though the coordinate system is quite appealing with these rainbow colors, I really doubt that adults in my age and with lower education level* *are able to comprehend it properly". (PN 1, male, 74 years).* In contrast, participants were able to associate the tachometer to real world artefacts, which helped them to understand the meaning of it. *"I like this arch. I don't know what this [coordinate system] is supposed to be though. [...]. Red, yellow, green...It is like a traffic light, you can associate it immediately. Green is ok, red is not and mediocre is yellow...tells me to pay attention. I would enjoy providing my data to see how it [tachometer] changes." (PN 4, female, 74 years).*

9.6.2   Intelligibility

Participants struggled to comprehend specific terms used within the prototype version of MY-AHA. The problematic terms were *interventions, cognition, app, emotions, exergames, export* and *share.* In general, participants mentioned that these terms were too scientific and technical. For PN 7 these terms are not used frequently in everyday language and therefore are hard to comprehend. The scientific, technical or unknown character of used terms had a deterrent effect on participants, as PN 9 phrased it, *"It is so scientific, so unbelievably scientific...It deters me, as I don't know much about such [scientific] things." (PN 9, female, 86 years).* Further, many participants did not comprehend the relation between MY-AHA as an integrative platform and the different third party solutions like various devices, interventions or applications. Explaining this relation to PN 6 did not result in a better understanding and resulted instead in increased anxiety about the complexity of the system. Even though other participants understood the meaning of interventions, they were not able to grasp the relation between these interventions and third party solutions, for instance the relation between a fall prevention intervention and the corresponding third party provider for the intervention system called iStoppFalls.

9.6.3   Usefulness

Participants' perception of the prototypes' usefulness was mainly characterized by three aspects, (1) summary of results within the dashboard screen, (2) provision of proactive system messages, and (3) information about individual health risks.

1. Dashboard: With exception of PN 1, all participants appreciated a summary of results through the dashboard, as it helped them to get a rough picture of their performance. However, they wished to be able to adjust the information presented on the dashboard. In this context, PN 5 said, *"The dashboard is valuable to me, as I get an overview of my performance. But I would like to change priority and order of the information shown". (PN 5, male, 78 years).* PN 7 and PN 8 expressed their desire for additional information on the dashboard. *"I would like* *to see my performance of already completed days. Now, I can only see my performance for the current day. But what about yesterday? I don't check my results for the day in the evening. I check them in the morning, but with this version of the dashboard, I cannot see these results. (PN 8, male, 65 years).* In contrast to the positive feedback, PN 1 questioned the value of the dashboard. *"The summary presented on the dashboard is incomplete. I want to see the detailed results anyway, so the dashboard does not provide any value to me". (PN 1, male, 74 years).*

2. Proactive system messages: Related to the dashboard, proactive system messages like notifications and suggestions with respect to the user's performance and results were perceived as valuable and helpful by most participants. In addition, the possibility of deactivating such messages was much appreciated. *"Well, I can deactivate them [notifications and suggestions] if I don't want to see them anymore. [...] But personally, I would leave them activated. I think they are helpful to stay at it". (PN 8, male, 65 years).* Nonetheless, some participants disliked proactive reminders by the system. *"Those reminders, I really don't like them. They annoy me. I would probably turn them off." (PN 1, male, 74 years).* Further, participants indicated that the different types of proactive messages by the system were not always comprehensible. The visual design did not clearly distinguish between the different types, which led to misunderstandings and misinterpretations. *"The difference between suggestions and feedback for instance is not distinct and I believe they belong together anyway. I would suggest merging them into one group". (PN 5, male, 78 years).*

3. Health risks: In general, participants ranked the provision of individual health risks through MY-AHA as useful information. With respect to risk simulation, participants perceived this feature as reasonable and motivational. In this context, PN 7 stated, *"Trying to reach the set reference values really motivates me, provided these values are not too ambitious". (PN 7, male, 68 years).* With a stronger focus on the reference values, PN 8 said, *"The reference values are reasonable. They constitute a good factor for orientation during training sessions". (PN 8, male, 65 years).* However, the presentation of risk values within MY-AHA caused misunderstandings and misinterpretations by users. A major problem here was the labelling of headings. For some participants, the meaning of the heading *activity risk* was not clear. *"[...] activity risk, what could that mean? Do I need to walk 10000 steps to reduce my risk here? Does it relate to fall risk? Do I need to gain muscles to affect it? Also what does that number mean, 1.5? Is this good? I really struggle to understand the meaning of it". (PN 4, female, 74 years).* Further problems concerned the calculation of risks, as the information provided was insufficient and led to irritation in participants. PN 1 assumed that the risk is calculated based on age only. PN 6 and PN 7 thought that they mistakenly manipulated the risk by interacting with interactive elements. PN 8 assumed that the risk was calculated on a daily basis.

9.6.4   Sustainability

Attempting to anticipate factors influencing sustainable use of MY-AHA by older adults, three topics emerged, (1) data security, (2) goals, and (3) Self-Assessment and integration into daily routines.

1. Data security: Willingness of our participants to use a system like MY-AHA is strongly correlated with data security aspects. PN 1 and PN 7 expressed their concerns during the registration process. *"Could you tell me which company operates this application? More importantly, would you tell me where my data will be stored? Will it be stored inside or outside of Germany?" (PN 1, male, 74 years).* In general, most participants were very reluctant to enter their chronic disease information into MY-AHA. With respect to textual explanations regarding data processing and protection, PN 9 remarked, *"I do not understand the explanations [regarding data security] provided here. I understand too little of such things, and a text like this cannot ease my concerns. [...] Probably I am so reluctant due to war experiences in my youth. You just weren't supposed to spread information easily back then (PN 9, female, 86 years).* In the context of data transfer to other parties, PN 5 and PN 9 mentioned their concerns for possible negative consequences of sharing health data, for instance increased health insurance contributions or disadvantages for family descendants.

2. Goals: Most participants welcomed the opportunity to set individual goals, because they saw it as positively affecting their motivation. Apart from this, some participants started to elaborate on additional goals, which might be implemented into MY-AHA. PN 4, PN 5 and PN 6 emphasized the importance of goals for the social domain. They suggested that MY-AHA should allow setting goals for e.g. volunteer activities measured in hours or speaking time measured in minutes per day.

3. Self-Assessment and integration into daily routines: Besides functionalities available and testable in the MY-AHA prototype version, participants speculated on their individual willingness to use self-assessment technologies on a daily basis, when integrated into full functional health platforms like MY-AHA. In this context, PN 4 could imagine a benefit from using self-assessment technologies with MY-AHA. *"I would try out such [self-assessment] technologies with a platform like MY-AHA. [...]. I could imagine using this [MY-AHA] on a daily basis". (PN 4, female, 74 years).* PN 5 shares the perspective of PN 4, however it seems discipline and social support are relevant for successful integration into daily routines. *"I can imagine using such a system [MY-AHA], but I would like to* *remark that this would require a lot of discipline. I would need my social environment to support me with that". (PN 5, male, 78 years).* Further, participants expressed their doubts that not all daily activities are considered in MY-AHA risk calculations and that they would need to make adaptations to their daily routines in order to receive an appropriate risk calculation. *"I am not sure if the system tracks for instance my daily gardening activities, or when I go swimming or take a nap. I am worried, that I need to change my daily routine just in order to receive accurate results". (PN 1, male, 74 years).*

9.7   Discussion

Technologies for AHA support clearly have the potential to improve health in older adults, as the existing literature demonstrates (Gschwind et al., 2015; Knight et al., 2014; Thomas and Bond, 2014; Vaziri et al., 2017; Verwey et al., 2014). Two issues as yet, however, have not been fully resolved. Firstly, it is not entirely clear that existing technologies are put to sustained use by older populations. Positive effects can only be achieved by means of sufficient exercise dosage and sustainable training with a good adherence over a longer period of time (Phillips et al., 2004), and thus the motivation and practices (Wulf et al., 2015c) of the participants play an important role in sustainability. Secondly, as yet, there are few examples of integrated health monitoring and assistive technology which address the many and disparate areas of concern for older persons' wellbeing. Given the widely varying nature of older people's needs, variation characterised in the personas we outline above, it is apparent that some fairly radical configurability is needed. This, in turn, needs to orient to the four broad design challenges we have discussed. Existing systems, we have argued, mainly focus on narrow and unidimensional perspectives on frailty, failing to deal adequately with the way in which different areas of health concern might interact with each other. Integrative health platforms constitute a promising approach to combining different AHA technologies and intervention programs and adjusting them to the needs and preferences of heterogeneous older adults to support them in AHA. However, ICT-based multidimensional solutions are novel and designing integrative health platforms appropriate for older adults remains a challenge. In our participatory design case study we developed a high-fidelity prototype, and accordingly, results in terms of prototype evaluation and design implications will be discussed in the following sections.

9.7.1   Addressing design challenges

The design challenges emerged from our pre-study have been considered relevant for older adults when engaging with integrative health platforms in previous research (Chaudhry et al., 2016; Eisma et al., 2003; Lindsay et al., 2012; Uzor and

Baillie, 2013). Therefore, we suggest they constitute a reliable frame to discuss our findings made when attempting to address these challenges in a prototype version of MY-AHA.

Improving usability: The application structure provided for both smartphone and tablet provided for an adequate understanding for most participants. However, we observed problems with respect to navigating through the application in older adults aged 80 years and above. As the usability tests demonstrated, there seemed to be a considerable learning overhead for these participants, for instance in terms of using mobile applications for the first time or solving the given tasks. Therefore, we believe that while the navigation structure seems to meet requirements of this target group, the applied test environment in particular was inappropriate for our participants, due to long duration and exhaustion. Our usability tests suggest that the benefits of orientation derived from applying a flat menu structure outweigh the drawbacks related to navigation in older participants (Carmien and Manzanares, 2014; Gudur et al., 2013; Lorenz and Oppermann, 2009; Torun et al., 2012).

In terms of icon design, we deliberately did not use icons with text labels in some screens of the prototype, mainly for reasons of insufficient space on the smartphone version. Here, we anticipated that participants would learn the meaning of icons when presented together with text labels, so they would recall them subsequently on other screens without text labels. However, such learning effects did not always occur in our study. According to the literature, skills and abilities in older adults to learn using new technologies are generally lower than for younger age groups (Marquié et al., 2002). Availability of training opportunities is one component restricting opportunities to learn new technologies in older adults (Rogers et al., 1996) and there was clearly an issue in our chosen test environment as to whether training opportunities for the target group were adequate.

For the visualization of health data, participants preferred a percentage display to a smiley (see fig. 6, right picture) and a curve representation to a bar diagram (see fig. 6, left and middle picture). Doyle et al. (2014) found similar results in their study, confirming that older adults may find curve representations helpful (Doyle et al., 2014). However, with respect to risk visualization participants preferred the tachometer with traffic light colors to the curve representation (see fig. 7). This is in contrast to other research implying that older adults struggle to comprehend visualizations that use traffic light colors (Doyle et al., 2014; Lorenz and Oppermann, 2009). The cited studies applied coloring backgrounds, while we purposely used strong-colored elements placed on white background. We assume that the different application of traffic light colors explains our distinctive findings.

Considering intelligibility differences: The prototype evaluation illustrated that participants had problems understanding specific terms, used to describe elements within the MY-AHA application. These terms either were of technical nature or used a foreign language (English). Arnhold et al. (2014) had previously confirmed issues of comprehension for older adults in their study (Arnhold et al., 2014). In addition, participants in our study also struggled to understand terms like cognition or intervention, which are not otherwise normally classified as technical terms or foreign language but are used less frequently in everyday language. It seems that older adults expect technologies to use terms related to their everyday context. This seems particularly true for health monitoring or self-assessment technologies, as health responsibility devolves from professionals to patients. The use of specialized terminology requires older adults to become more familiar with, for instance, measurement outcomes which are not otherwise easily understood, and therefore constitutes an additional barrier for them.

Enhancing usefulness for the target group: Older adults in our study perceived the summaries of results presented on the dashboard screen as useful. In addition, proactive system messages seemed to enhance perceived usefulness on the one hand, but also showed to have the potential to promote reluctance to use AHA technologies in some older adults. The latter seems especially true for reminder messages, which were explicitly disliked by some participants. Smith et al. (2016) suggest that such messages need to address the individual preferences and personalities of end users (Smith et al., 2016). In this context, other studies illustrate that presentation, frequency and content are individual factors determining whether proactive system messages are perceived as useful or annoying (Bailey et al., 2001; Goldstein et al., 2014; Haberer et al., 2012). Further, our study showed that the textual design and wording of proactive system messages might affect participants' perception of such functionalities' usefulness. In general, our study indicated that configurability of the system seems to be a crucial aspect for older adults' perceived usefulness of AHA technologies. Such findings are emphasized by scientific work from Procter et al. who argue for more customizability and configurability in AHA technologies (Procter et al., 2014) and Wulf et al. who elaborated on flexible applications that allow end users to tailor software to their needs (Wulf et al., 2008). The design of AHA technologies may build upon such work by transferring ideas and concepts of configurability into an ICT supported healthcare context.

Promoting sustainable technology use: Participants were very concerned with data security and the possible consequences of data transfer to third parties. Throughout the usability tests, they expressed their fears and concerns by themselves, without having the moderator initiate data security topics. Most of the concerns were consistent with our pre-study findings, where we explicitly asked for participants' opinions. The fact that they mentioned most of them again during the usability tests underlines the significance of data security for our participants. The relevance of data security in the context of AHA technology use by older adults is addressed in many other studies (Coughlin et al., 2007; Latulipe et al., 2015; Publications Office of the European Union, 2012; Schulz et al., 2015). According to our results and the respective literature, older adults require time to establish a foundation of trust in AHA technologies. This means that sensitive data, for instance older adults' chronic diseases, should not be requested early on. Studies show that older adults' willingness to share sensitive data significantly depends on trust in technology (Hernández-Encuentra et al., 2009; Wilkowska and Ziefle, 2009).

Finally, our results indicate that integration of MY-AHA into daily routines of older adults seems to influence their willingness to use such technologies long-term. Participants in our study were concerned about the integrability of MY-AHA into their daily life, as they doubted that specific daily activities like gardening or swimming would be tracked and considered for risk calculation. Participants feared that they would need to change daily routines in order to receive an accurate risk calculation. In fact, such concerns are echoed by other studies investigating the use of self-assessment technologies in older adults (Hänsel et al., 2015; Schlomann et al., 2016). In their studies, they conclude that integration of self-assessment technologies into individual daily life routines is key for long-term acceptance by older adults. Our results strongly suggest the same. Apart from daily life integration of AHA technologies, participants emphasized their need for social support in order to embed prevention programs in their daily life successfully. They defined social support typically in terms of accompaniment in prevention activities and programs. Literature on persuasive design uses social support as an approved concept in motivating older adults (Halko and Kientz, 2010; Torning and Oinas-Kukkonen, 2009).

9.7.2 Design implications for integrative health platforms

Based on the findings presented in this study, we were able to derive eight implications, relevant for the design of integrative health platforms like MY-AHA, (1) build trust, (2) improve comprehensibility, (3) raise awareness, (4) apply an individual persuasion strategy, (5) strengthen autonomy, (6) support individual socio-technical fit, (7) improve visual accessibility and (8) enhance learnability. While our study focused on an integrative health platform, the implications might as well be valid for single AHA technologies such as activity monitors, dietary coaches or sleep monitors.

Build trust: Integrative health platforms should allow the user to familiarize with the approach of digital collection and storage of health data. Therefore, sensitive health data should not be requested early on, for instance during the registration process. Older adults should be allowed to try out the application and gain practical experiences in order to reassure them that the system will prove useful over time and to alleviate any concerns over trust, privacy and so on. Further, the visual appearance of the application needs to represent an appropriate degree of seriousness and reliability. This may be achieved by providing clear and logical structures, a consistent appearance and a synchronized color scheme. Finally, the application needs to provide sufficient information about the operator of the application and transparency about the procedures of data processing. According to our study, it seems that older adults with high level of experience with mobile technology get most benefit from information and transparency, while older adults with low technology experience benefit more from time and visual aspects mentioned in this section. Such trust building attitudes are especially relevant for the users we identified as *skeptical users*.

Improve comprehensibility: The graphical user interface, as we have stated, should avoid scientific, medical or technical terms, as well as terms in foreign languages. Instead, the application might use terms related to the everyday language of older adults. In case specialized terms cannot be avoided, explanations appropriate for the target group have to be provided. For comparison and interpretation of individual health data, the use of individual health reference values may counteract false expectations or overexertion induced by comparisons between individual health data and general reference values, derived from professionals or from the literature. Depending on the condition, general reference values may fail to serve as a reasonable and appropriate orientation. In this context, proactive system messages need to consider an adequate and sensitive wording, especially when measurements exceed the normal range, to prevent negative consequences on the emotional well-being of older adults. Further, proactive system messages should include suggestions or recommendations directly convertible into everyday life. In terms of health data and results presentation, visualizations need to be as simple as possible. The key message of a chart should be comprehensible at first sight. Therefore, forms of presentation familiar to the target group work well. In our study, such forms were curve diagrams and tachometers. Finally, color-coding can considerably improve comprehensibility of displayed results and information. For instance, older adults classify traffic-light color schemes, which are well known in everyday life, as understandable. Improving comprehensibility of AHA technologies, will ease concerns and demands of *worried users*.

Raise awareness: Illustrating health benefits by means of reasonable and transparent measures and effects of prevention may increase older adults' awareness with respect to active healthy ageing. Practical experiences seem to be key for older adults to perceive prevention activities as useful. Providing experience reports or recommendations of peers might help attract older adults' attention for prevention activities. Such functionalities therefore can constitute a valuable component in the design of integrative health platforms like MY-AHA and raise awareness in *interested and needy users.*

Apply an individual persuasion strategy: Persuasive elements such as proactive feedback, goals or reminders should only be used by one's own choice. Allowing configuration of these functionalities according to the users' preferences is crucial to address concerns and fears with respect to controllability. If support through persuasive elements is desired, proactive messages and goals need to behave in a context-sensitive manner, appropriate to measured values and user behavior. Persuasive elements might affect *performance-minded* and *skeptical users* the most.

Strengthen autonomy: A modular structure of prevention programs and alternatives for each health domain provides older adults with the autonomy to select preventive measures suitable for their individual life style, demands and interests. This might be especially relevant for *performance-minded users,* who want to follow an individual prevention plan that matches their other activities and *worried users* who fear to lose autonomy by using AHA technologies. Further, the wording of proactive messages needs to address individual characteristics and preferences of older adults to prevent feelings of patronization or anxiety, for instance in *worried users.* Moreover, proactive messages should always follow a rationale that is easily understood by older adults. Functionalities with respect to data transfer to other parties within the healthcare system should be transparent and their execution should remain in the hands of older adults at all time, to ease concerns of *skeptical users.*

Support individual socio-technical fit: Self-assessment technologies, as well as integrated health platforms like MY-AHA are likely to collide with already existent practices of people. Thus, there are health-promoting activities, for instance swimming or dancing, which older adults integrated into their daily routines, long before AHA technologies entered the scene. AHA technologies are not capable to monitor such activities reliably. To prevent frustration in older adults, *performance-minded* and *needy users* in particular, easy and effortless alternatives to enter such data manually need to be provided.

Improve visual accessibility: Consideration of accessibility standards are mandatory, when designing AHA technologies for older adults. Relevant aspects refer to findability and visibility of functions, sufficient color contrasts, highlighting important elements through coloring and finding a balance between font and display size. Such aspects might positively affect *interested* and *needy users.*

Enhance learnability: The design of an integrated health application like MY-AHA should probably apply a standardized design framework, familiar to the target group, to reduce learning efforts. Context-sensitive clues may support the user to find specific functionalities or areas within the application. In this context, older adults who are entirely new to mobile technologies could receive a tutorial on how to use touch screen gestures to support findability and reduce required efforts. However, such affordances need to be carefully applied as older adults (like everyone else) can perceive them as annoying later on, particularly if they feel they are being patronized. *Interested* and *needy users*, who either require technical support or have a very low tolerance threshold for technical issues, might benefit most from aspects enhancing learnability.

9.8   Limitations

The results presented in this article have to be considered in the light of limitations with respect to the pre-study, the prototype and its evaluation. Our pre-study only included one participant of advanced age (85 years), while the remaining participants were aged between 65 and 78 years. Further, many participants in our study were members of a senior organization, we collaborated closely with and a considerable proportion of those members, who were included in our study, were well educated, socially connected and aware of the significance of prevention for their own health. Participants in our sample therefore do not represent the whole spectrum of older adults and their diversities. No study, we would suggest, can. It is more important that we become sensitive to issues of heterogeneity and learn to design for it. Moreover, functionality of the high-fidelity prototype was restricted to certain use cases, which is why some dialog boxes or dynamic modifications triggered by user interaction were not operating fully. Auditory aspects, for instance spoken instructions for interventions, could not be evaluated, as they were not included in the prototype. Further, the prototype only presented a screenshot of health data collected during the pre-study. It is to be expected that user behavior differs when dealing with static data presentations instead of dynamic health data. At last, our evaluation setup only provided a snapshot view on user perception and interaction with the prototype in a laboratory setting and therefore disregarded learning and appropriation effects. Such effects in older adults only uncover fully when engaging with technological artefacts over long and in their daily life contexts.

9.9   Conclusion

In this article, we conducted a participatory design study for an integrative health platform named MY-AHA. Based on pre-study findings, we identified relevant design challenges and attempted to address them in a high-fidelity prototype. The prototype then was tested with the target group to derive implications and challenges for the further design of MY-AHA. Procedures and results documented in this design case study illustrate and explain how we made design decisions and how we addressed design challenges in the prototype. A recurring rationale throughout this article has been the fact that configurability of AHA technologies seems to be a crucial aspect when designing for older adults, as it supports heterogeneity of this target group more appropriately. Considering the heterogeneity of older adults seems to be a key factor when aiming for sustainable AHA technology use. Our study contributed to the understanding of heterogeneous demands and needs of older adults and provided an exemplarily approach to address these in the design of technologies for AHA support. Even though we were not able to illustrate any long-term effects for MY-AHA at this early stage of development, what we can say is that designers of AHA technologies need to consider as many opportunities for sustainable use as possible in their designs to promote long-term technology use by older adults. To identify such opportunities, design case studies may function as an appropriate framework for guidance. In the future, we will carry out a long-term evaluation study with a more mature version of MY-AHA to investigate long-term use, motivation and integration into older adults' real-world contexts and routines.

10 Conclusions

Technologies for AHA support can be a viable alternative to promote health in older adults. They provide functionalities to support a healthy lifestyle in daily life, for instance by monitoring older adults' activity and suggest prevention or intervention activities. Field studies illustrate that the use of AHA technologies for AHA support may effectively improve or maintain health in older adults (Finkelstein et al., 2016; Gschwind et al., 2015; Mansi et al., 2015). However, a major requirement for health improvements in these studies is long-term use of such technologies by older adults. Even though, research investigating influencing aspects of long-term usage by older adults provides a better understanding (Ballegaard et al., 2008; Martin et al., 2000), appropriately addressing these aspects in the design of AHA technologies to facilitate daily life integration and promote sustainable long-term use remains a major challenge (Di Pasquale et al., 2013; Fitzpatrick and Ellingsen, 2013; Jarman, 2014; Ogonowski et al., 2016; Wan et al., 2016). In order to contribute to the scientific discourse, this thesis illustrated the process of investigating older adults' practices and attitudes with respect to health, quality of life and AHA technology use and suggested how to address them in the design of technologies for AHA support to facilitate integration into older adults' daily lives and thus promote sustainable long-term use and health impacts. Therefore, empirical evidences were collected by carrying out mixed methods analyses of practices and attitudes of older adults and their related stakeholders on health, quality of life and AHA technology use. Based on these analyses, a PD case study defined specific challenges for the design of technologies for AHA support in older adults and addressed them in an AHA technology prototype. Finally, the prototype design was evaluated with older adults and implications for future design were derived.

10.1 Summary of findings

10.1.1 Relevance of heterogeneity in the population of older adults for AHA technology design

In general, technical aspects of ICT-based fall prevention systems like usability or user experience were relevant factors for older adults' technology acceptance. In this regard, the study suggested that malfunctions, complex visualizations and obtrusive design were factors impeding older adults' willingness to use ICT-based fall prevention systems like iStoppFalls. The perception of these factors seemed to be influenced by participants' gender and age. In this context, provided games of iStoppFalls became boring for male participants after some time. According to

© Springer Fachmedien Wiesbaden GmbH, part of Springer Nature 2018
D. D. Vaziri, *Facilitating Daily Life Integration of Technologies for Active and Healthy Aging*, Informationsmanagement in Theorie und Praxis,
https://doi.org/10.1007/978-3-658-22875-0_10

the interviews, many male participants experienced the exergames to be too repetitious and wished for a progression of the difficulty level to make them more interesting. In contrast, most female participants enjoyed playing the games and did not express their desire for a more challenging experience. Therefore, the author's study suggest that male participants seek for challenges when using ICT-based fall prevention systems like iStoppFalls, while female participants prefer an easy experience. With respect to age, the study indicated that younger participants were more critical with respect to graphics and appearance of exergames than older participants were. Other research investigating usability and user experience aspects in older adults' technology use show similar results, for instance in (Kari et al., 2012; Phiriyapokanon, 2011; Sheikh and Abbas, 2015; UNHCR, 2015). Furthermore, older adults in the study were positively attracted by the applied activity monitor (SMM) and its functionalities to track and collect data about their activity. This is an interesting finding, as literature frequently underlines older adults' reluctance to technologies that collect health data (Braun, 2013; Heart and Kalderon, 2013; Lee et al., 2013; Miller and Bell, 2012; Morris and Venkatesh, 2000). In the author's study however, such functionalities motivated older adults to use the iStoppFalls system and engage in physical activity. The author believes that the research context in general and further the frequent contact between researchers and participants established a foundation of trust and thus eased older adults' skepticism and fear with respect to data privacy. Such motivating factors of health data collection and monitoring are in line with results from a subsequently conducted case study presented in chapter 6 of this thesis, which investigated the effects of the iStoppFalls system on subgroups of older adults and analyzed their practices and attitudes with respect to system usage. The study suggested that older adults, who used the SMM and iStoppFalls exergames more frequently, were able to reduce their fall risk significantly compared to older adults who used either of those less frequently. From interviews with study participants, the author learned that participants perceived wearable devices as an additional motivator to engage in physical activity. Initially, the SMM was implemented in the study for mere data collection. The author and collaborating researchers did not anticipate that participants would attribute the SMM with an active role. This emphasizes the challenge of researchers in this field to think about all possible scenarios of how participants may use technology. In this context, examination of older adults' practices provided the author with useful additional data with respect to AHA technologies. Participants expressly underlined that the possibility to use different AHA technologies better accounts for their heterogeneous practices and attitudes and thus facilitates their motivation to use such systems in daily life. Referring to Pelle Ehn (1990), this example illustrates that a design artefact does not define by the meaning we as designers attribute, but by what the artefact means to the user in practical use and understanding (Ehn, 1990). Furthermore, interviews with older adults implied that social aspects like being active together with others or receiving family support were important factors in their motivation to engage with ICT-based fall prevention systems.

Based on these findings, the author concludes that an ICT-based fall prevention system like iStoppFalls has the potential to be an appropriate AHA technology for older adults. However, as participants articulated during the interviews, the focus on a single technology does not account for the heterogeneous practices and attitudes in the population of older adults and AHA technologies like iStoppFalls need to provide opportunities to involve older adults' social environment in order to facilitate AHA technology use.

10.1.2 Influence of older adults' social environment and related secondary stakeholders on their AHA technology use

Existing literature mainly understand technologies for AHA support to be a mediator for older adults' engagement in healthy activities (Graham et al., 2014; Rimmer et al., 2004). In contrast, the author's study illustrate that the opposite might be true as well. During the research process, the author learned that older adults' engagement in a healthy lifestyle may have mediating effects on their use of AHA technologies and that older adults' social environment seems to take considerable influence here. In interviews, participants frequently mentioned social support from family and friends, social participation and being active together with peers, for instance in group training classes, as relevant factors for them to engage in healthy activities. These findings and corresponding literature suggest that the involvement of older adults' social environment might facilitate the integration of healthy activities into older adults' daily lives (Cialdini, 2001; Goldberg and King, 2007; Guedes et al., 2012; Hall et al., 2010; Kahn et al., 2002; Locke and Latham, 2002; Thraen-Borowski et al., 2013). Therefore, it seems that the integration of healthy activities, as well as the integration of technologies for AHA support into older adults' daily lives constitute two relevant challenges in health technology design. Presented results in this thesis imply that both challenges may be addressed by involving older adults' social environment into AHA activities and thus the design of AHA technologies should focus on concepts and functionalities that facilitate the involvement of older adults' family members, friends and acquaintances. These results strengthen previous findings of the author's study presented in chapters 5 and 6 of this thesis. However, the results only concerned with older adults' circle of family, friends and acquaintances and ignored interactions and relations to other stakeholders that might be of relevance here. The conducted case study in chapter 8 suggested that secondary stakeholders in the healthcare system may considerably influence older adults' practices and attitudes with respect to a healthy lifestyle and technology use for AHA support. In this context, the author learned that while older adults and secondary stakeholders do agree in some regards, for instance the need for data security and privacy, they have considerably contradictive perspectives and conceptions with respect to independence, well-being and trust. In terms of independence, it seemed that especially doctors were worried about older adults improving their health literacy. They were concerned about increased resistance to medical advice in older adults. In contrast, older adults' motivation to increase their health literacy was mainly driven by their wish to understand check-up procedures and their outcomes. They understood health literacy as a possibility to shift from passive receipt of medical instruction to active involvement in own health. This finding coincides with research results of Lorenzen-Huber et al (2011) who found that older people do not wish to be passively monitored subjects, but desire to be treated as equals (Lorenzen-Huber et al., 2011). With respect to well-being, secondary stakeholders mainly associated well-being with measurable physical and cognitive health and tend to follow a deficit approach here. In contrast, older adults associated well-being with enjoyable activities, even if these were unhealthy. Finally, the case study implied that older adults have a considerable distrust in health insurance companies to exploit collected health data to older adults' disadvantage. Such contradictions seem to promote reluctance of older adults to use AHA technologies in their daily lives. Hence, in order to facilitate daily life integration of AHA technologies, health technology design in this context needs to understand and negotiate different perspectives and conceptions of older adults and secondary stakeholders by involving these parties in the design process.

10.1.3 Designing AHA technologies to create opportunities for daily life integration

Previous research presented here implied that heterogeneous practices and attitudes of older adults and their interactions and relations with their social environment and related secondary stakeholders are key elements that need to be addressed in the design of technologies for AHA support in order to facilitate their integration into older adults' daily lives. Based on these findings, the author designed and evaluated a prototype for an integrative health platform in a participatory design case study with older adults. The platform at its current state allows older adults to simulate the combination of different AHA technologies that suit their practices and attitudes and the processing of collected data of the devices with respect to their health condition and preferences for prevention programs. At this stage, functionalities to communicate and exchange information with other stakeholders like relatives, doctors or health insurance companies were not included in the prototype.

The design case study contextualized heterogeneous practices and attitudes of older adults into specific design challenges and illustrated how these may be addressed in the design to create opportunities for the integration of AHA technologies into older adults' daily lives with regard to different health domains in need, e.g. physical activity, cognition, nutrition, sleep, mood and social connectedness. In this context, the author defined usability, intelligibility, usefulness and sustainability as relevant challenges for the design of AHA technologies. Usability mainly referred to reliability aspects that facilitate trust building of older adults. Here, older adults in the author's study wished for a consistent design and predictable system behavior. According to existing literature, such aspects seem to be factors that support trust building in older adults (Hernández-Encuentra et al., 2009; Mendoza, 2009; Minnick, 2016; Wilkowska and Ziefle, 2009). Therefore, the prototype oriented towards known design guidelines and appearances with a low level of complexity. In terms of intelligibility, older adults' low levels of experiences with mobile and ubiquitous technologies request didactic usage support, for instance learning to use the system through a step-by-step approach or by appropriate instructions. Here, the system identified context sensitive clues as an appropriate instrument to facilitate older adults' interaction with the platform. Further, the author's study and related literature suggest that the design of technologies for AHA should carefully consider used terminology, as technical or foreign language terms generally are hard to comprehend by that target group (Arnhold et al., 2014). The author identified relevant technical and foreign terms like *cognition* or *intervention* during the study and selected alternative terms together with participants. Usefulness of the system seemed to be mainly determined by older adults' opportunities to choose prevention programs and associated AHA technologies on their own, and the health information provided by the system, which, according to older adults in the study, should only be on the purpose or operation of prevention programs. In this context, older adults' perception of a technology's usefulness seemed to be influenced by the notifications, suggestions or feedback provided by the system. Proactive support may be welcomed by some participants, but other participants may feel patronized and stop using the AHA technologies. Therefore, health technology design needs to address individual preferences in the population of older adults with respect to such functionalities. Existing literature suggests that presentation, frequency and content of proactive system messages constitute relevant features that need to be adjusted to heterogeneous requirements of older adults (Bailey et al., 2001; Goldstein et al., 2014; Haberer et al., 2012; Smith et al., 2016). Here, one key implication derived from the presented case studies in this thesis is the apparent need for a higher degree of configurability in technologies that support AHA of older adults. To address such demands the platform allowed the user to set individual settings for notifications and reminders.

With respect to sustainability of AHA technologies, participants articulated the importance of seamless integration into their daily life activities, if they were to use AHA technologies long-term. In this regard, three major aspects for older adults in the case study were the compatibility of AHA technologies with other mobile technologies already in use, the possibility of providing controlled health data access to trusted doctors for the purpose of diagnostics and the unobtrusive and easy use of technologies for AHA support to reduce daily effort. Participants raised these points, mainly in the light of concerns that they would need to change their daily routines in order to make use of technologies for AHA support. Other studies underline these findings and thus this thesis strongly suggests to consider aspects of daily life integration carefully in the design of AHA technologies (Hänsel et al., 2015; Schlomann et al., 2016).

Finally, participants in the design case study emphasized their desire for social support. In general, they defined social support in terms of accompaniment in healthy activities and prevention programs. These findings underline the implications derived from the other case studies presented in chapters 7 and 8 and further are supported by design literature that uses social support as an approved concept in motivating older adults (Halko and Kientz, 2010; Torning and Oinas-Kukkonen, 2009).

10.2   Research limitations

The results presented in this thesis contribute to the scientific discourse and provide insights for the design of AHA technologies that aim at long-term use and sustainable impacts on older adults' health. However, it is important in this context to inform about limitations and possible shortcomings of the presented research, to put the results into perspective.

The idea for the design of an integrative health platform is partly based on results of the iStoppFalls project. These results were generated in a long-term study over 6 months. While classifying a 6 months study as long-term may be debatable in itself, the results were only able to represent practices and attitudes of older adults' technology use with respect to the summer season of the year. This is problematic, as technology use in winter season may considerably differ from summer season and vice versa. Therefore, the author is aware of the fact that the design of the integrative health platform may have missed relevant practices and attitudes of older adults that would only have revealed in winter season and may have differed from those investigated during the summer. According to corresponding literature, this is likely to be true, as seasonal mood is evidently proven (Lansdowne and Provost, 1998; Murray, 2003). Hence, the living lab and RCT for the MY-AHA project spans 18 months to allow for a more detailed investigation of longer AHA technology use of older adults and take account of effects like seasonal mood.

Besides such season-related aspects, studies that investigate populations across different countries face challenges of cultural diversities and thus different practices and attitudes of respective older adults become relevant for the design of AHA technologies. The presented case studies in this thesis did not explicitly consider such differences of community dwelling older adults living across European countries and Australia. Hence, practices and attitudes were not investigated in the light of their respective cultural background and how this may affect the design of AHA technologies. However, the design of AHA technologies needs to consider such vantage points as well in order to facilitate daily life integration of such technologies and promote their sustainable implementation in different healthcare systems across and outside Europe. Future research in the MY-AHA living lab will provide a space for rigorous investigation of cultural aspects across different countries and will put them into perspective for health technology design.

Further, this thesis emphasized the importance to involve not only older adults but also other relevant stakeholders in the design process of AHA technologies. The research presented in the case studies did only report on perspectives of other stakeholders and their interactions with older adults, but did not bring all parties physically together, for instance in design workshops. Such meetings could have generated additional insights into relations and interactions between older adults and relevant stakeholders and thus more specific implications for the design of the integrative health platform prototype may have been derived. In this context, the prototype has been evaluated with older adults only and excluded other relevant stakeholders for the time being. The main reason for this procedure is the early stage of the prototype and its limited functionality. Functions that allow collaboration and cooperation between older adults and relevant stakeholders were not integrated yet. In a next prototype version such functionalities will be addressed and allow the evaluation of interactions between older adults and relevant stakeholders that are realized through the integrative health platform.

Finally, throughout the case studies older adults were visited at their homes by researchers, which is a key element of the living lab methodology. In many cases, that provided opportunities for conversations and the establishment of trust relationships that might have affected older adults' motivation to participate in the studies. Therefore, there is a possibility that results presented in this thesis were biased by such social interactions. For the design of AHA technologies this constitutes a considerable challenge, as social interactions and relationships with participants help researchers to gain a detailed understanding of older adults' practices, attitudes and perspectives and support the design of appropriate solutions for that target group. However, sustainable implementation of AHA technologies within the healthcare system might be compounded by the fact that these social interactions and relationships between researchers and participants cannot be maintained infinitely and their conclusion may affect older adults' motivation to further use AHA technologies after the project ends.

10.3 Discussion of the methodological approach

As the case studies showed, long-term use of AHA technologies by older adults constitutes a major challenge and thus sustainable health impacts induced by AHA technology use are rare. Part of the problem is the high degree of heterogeneity in the population of older adults and the complexity of interactions and relations between relevant stakeholders within the healthcare system. Such aspects, if not appropriately addressed by health technology design, exacerbate technology integration into older adults' daily lives and constrain opportunities for long-term use. The goal of this thesis was to investigate practices and attitudes of older adults in the context of AHA technology use and to illustrate how designers may address them in health technology design to facilitate the integration of technologies for AHA support into older adults' lives and thus create opportunities for long-term use. Results illustrate that the integration of multiple AHA technologies and the involvement of older adults' social environment seem to be key factors facilitating daily life integration of such technologies. Further, the thesis suggests that technologies for AHA support need to provide a certain degree of configurability in order to address heterogeneous requirements of older adults. However, the author's study implies that daily life integration of such technologies may not be facilitated by these factors alone. What seems to be closely linked to the integration of AHA technologies into older adults' lives is the possibility for relevant stakeholders like policy makers or health insurance companies to uptake AHA technologies and related research results and implement them into healthcare systems. From the case studies, the author learned that daily life integration may be achieved more seamlessly, if secondary stakeholders understand, accept and recommend technologies for AHA support in older adults. Here, it is important to convince these stakeholders of technologies' benefits and values for AHA support in older adults. In general, these parties are more interested in measurable information, for instance about user acceptance or system efficacy that help them to plan and anticipate the success of market entry strategies for sustainable impacts. Hence, an additional challenge for research in the field of AHA technology design for older adults is to provide lucid information for relevant stakeholders to ease the uptake of these technologies and research results and thus promote sustainable long-term use and health impacts.

In this light, the thesis systematically established a detailed understanding of older adults' practices and attitudes with respect to health, quality of life and technology use by using qualitative and quantitative methods. One may argue that qualitative methods already equip researchers with appropriate tools to investigate practices and attitudes systematically. In fact, results of qualitative research did provide relevant insights into specific cases and elucidated meanings of older adults' practices and actions. However, specification here may also increase the risk to overlook interactions and relations that might only reveal when investigating larger amounts of cases from a more abstract perspective (Brannen, 2005). Pulling back to see the bigger picture and putting it into perspective with specific cases might be a reasonable approach to provide more detailed answers to research questions in the context of health technology design for older adults.

With regard to this thesis, quantitative methods helped to reveal following interactions of older adults', which may have remained unnoticed if exclusively investigated by qualitative methods; (1) the decrease of older adults' fall risk affected by the combination of different AHA technologies, and (2) the engagement of older adults in healthy activities, as a significant factor for their willingness to use AHA technologies. Qualitative analyses then exemplified these interactions by providing and elucidating the meanings that older adults attribute to their practices and actions.

With respect to the first quantitative finding, qualitative interviews indicated that the combination of different AHA technologies seems to provide older adults with a wider range of usage scenarios, which allows them to integrate the system into their daily practices and routines more easily. Easy integration into daily practices came to be a key factor for older adults and affected their motivation to use and adhere the system more frequently, and it is likely that this led to a reduction in their fall risk. Literature illustrates that the use of ICT-based interventions and activity monitors positively affects older adults' health (de Bruin et al., 2008; Schoene et al., 2011; Smith et al., 2011) and that integration into daily life routines is important for long-term use and adherence (Ballegaard et al., 2008; Martin et al., 2000). However, research on combinational effects of AHA technologies on older adults' health and practices in this context is rare.

For the second quantitative finding, interviews indicated that older adults' engagement in healthy activities considerably depends on interactions and relations with and to their social environment. Self-determination, social participation, social support and having company were key aspects for older adults to engage in healthy activities. Such effects have been confirmed by previous research (Goldberg and King, 2007; Guedes et al., 2012; Kahn et al., 2002; Thraen-Borowski et al., 2013). However, considering qualitative and quantitative findings together, results suggested that healthy lifestyle might be a mediator for older adults' use of AHA technologies and that their social environment may have considerable influence here. Current literature mostly sees AHA technology use as a mediator for healthy activities in older adults and not vice versa (Graham et al., 2014; Rimmer et al., 2004). Based on these findings, a qualitative interview study with older adults and secondary stakeholders in the healthcare system elucidated the meanings of practices and interactions between older adults and their social environment.

As described above, the convergence of qualitative and quantitative methods generated more insight into older adults' practices and attitudes with respect to health, quality of life and AHA technology use and revealed unexpected results for further investigation, for instance the motivational effects of combining AHA technologies or the mediating role of healthy activities for AHA technology use by older adults and the influence of their social environment. On the one side, converging both methods may enable designers to address older adults' heterogeneity more accurately within their designs. On the other side, the convergence may help researchers in the context of AHA technologies to address their findings to a wider range of stakeholders in the healthcare system, for instance policy makers or health insurance companies who generally prefer measurable information. The uptake of technologies for AHA support by these parties plays a major part for sustainable long-term use of such technologies and health impacts in older adults and therefore, the ways relevant stakeholders situate and understand research findings need to be in the focus (Green et al., 2015). In fact, insufficient translation of research findings in the past led to an increasing demand from outside academia for more practical research to inform relevant stakeholders within the healthcare system and ease the uptake of research in the field of AHA technologies (Berwick, 2003; Hammersley, 2000; Proctor et al., 2009; Westfall et al., 2007). In this context, literature suggests that research must be designed, disseminated and implemented in concert with relevant stakeholders and therefore researchers need to learn about practices, attitudes, needs and perspectives of a full range of stakeholders, including e.g. policy makers, clinical staff or families of older adults (Green et al., 2015). Such challenges become especially true in research environments that include dissemination and implementation of research results into practice (Berwick, 2003; Hammersley, 2000; Proctor et al., 2009; Westfall et al., 2007). This was the case for conducted research in the context of iStoppFalls and will be paramount in the context of the upcoming MY-AHA research activities. Therefore, the author suggests that mixed methods designs may support the development of a deeper understanding of practices, attitudes and perspectives of involved stakeholders that affect adoption and implementation of AHA technologies into healthcare systems. Such research designs may address strengths of qualitative methods in providing an in depth understanding of older adults' practices, attitudes, perspectives and relations to other stakeholders within their social environment that are of major relevance at early project phases, for instance to design and evaluate appropriate prototypes with and for older adults. Further, the use of quantitative methods in such designs may provide extra perspectives on research objects and equip researchers with additional possibilities to translate results into appropriate information for other relevant stakeholders within the healthcare system and thus promote sustainable implementation and dissemination of AHA technologies. In general, this becomes much more important at late project phases (Green et al., 2015; Stetler et al., 2006). Therefore, both (1) the goal of the investigation and (2) the time or sequencing of qualitative and quantitative approaches may determine the way qualitative and quantitative data will be combined in health technology design (Bryman, 2012).

This thesis illustrated how the design of AHA technologies can create opportunities for long-term use by facilitating integration into older adults' lives. Future research may build upon these results and investigate how these opportunities affect older adults' use of integrative health platforms and affect sustainable improvements of their health and quality of life in long-term studies. In this regard, a remaining challenge for research is to define the duration of long-term studies in order to provide solid data on system use and health impacts. Such studies need to run long enough to provide reliable answers to relevant research questions, for instance adoption and acceptance of AHA technologies or efficacy of AHA technologies to improve or maintain older adults' health. In this context, literature suggests study setups that span several months (Anderson-Hanley et al., 2012; Gerling et al., 2015; Jung et al., 2009; Müller et al., 2015a; Ogonowski et al., 2016; Wan et al., 2014), although the investigation of sustainable long-term use and health impacts, would require researchers to monitor older adults' AHA technology use over several years or even longer in order to generate reliable data. However, the pace of technological progress may constrain the design of studies that last so long, as research results for state of the art AHA technologies may ease the uptake by relevant stakeholders. Such conflicts aggravate the definition of long-term in the context of health technology design and call for more emphasis on this topic by future research.

To conclude the methodological discussion, the use of qualitative methods in health technology design studies may provide rich and detailed answers to a wide spectrum of research questions relevant to understand technology integration into older adults' daily lives, effects on their health, their quality of life and their use of AHA technologies and further suggest implications for an appropriate design for that target group. However, reasonable convergence of qualitative and quantitative methods, as illustrated in this thesis, may allow additional perspectives on research results and improve translation for other relevant stakeholders within the healthcare system, which may ease the uptake of research results by these parties and support sustainable implementation of technologies for AHA support into the healthcare system. According to the author's study, both, possibilities for uptake and implementation seem to be relevant factors to facilitate the integration of AHA technologies into older adults' daily lives. Further, the combination of qualitative and quantitative methods may alleviate reservations and allegations of involved stakeholders with respect to limited generalizability of qualitative findings and limited depth of understanding of quantitative findings, which are both challenges that need to be addressed in health technology design. Thus, the use of mixed methods approaches may facilitate collaboration and cooperation of involved parties and improve the design of technologies for AHA support that aim for daily life integration, long-term use and sustainable health impacts of older adults.

References

Abowd, G.D., Bobick, A.F., Essa, I.A., Mynatt, E.D., Rogers, W.A., 2002. The aware home: A living laboratory for technologies for successful aging. Presented at the Proceedings of the AAAI-02 Workshop "Automation as Caregiver, pp. 1–7.

Active Ageing, 2002. A policy framework. World Health Organ. 59.

Agmon, M., Perry, C.K., Phelan, E., Demiris, G., Nguyen, H.Q., 2011. A Pilot Study of Wii Fit Exergames to Improve Balance in Older Adults: J. Geriatr. Phys. Ther. 1. doi:10.1519/JPT.0b013e3182191d98

Almirall, E., Wareham, J., 2008. Living Labs and open innovation: roles and applicability. Electron. J. Virtual Organ. Netw. 10, 21–46.

Amberg, M., Fischer, S., Schröder, M., 2005. An evaluation framework for the acceptance of web-based aptitude tests. Electron. J. Inf. Syst. Eval. 8, 151–158.

Amberg, M., Hirschmeier, M., Schobert, D., 2003. DART — Ein Ansatz zur Analyse und Evaluierung der Benutzerakzeptanz, in: Uhr, W., Esswein, W., Schoop, E. (Eds.), Wirtschaftsinformatik 2003/Band I. Physica-Verlag HD, Heidelberg, pp. 573–592.

Amireault, S., Godin, G., Vézina-Im, L.-A., 2013. Determinants of physical activity maintenance: a systematic review and meta-analyses. Health Psychol. Rev. 7, 55–91. doi:10.1080/17437199.2012.701060

Ammenwerth, E., Iller, C., Mahler, C., 2006. IT-adoption and the interaction of task, technology and individuals: a fit framework and a case study. BMC Med. Inform. Decis. Mak. 6, 3. doi:10.1186/1472-6947-6-3

Andersen, T., Bjørn, P., Kensing, F., Moll, J., 2011. Designing for collaborative interpretation in telemonitoring: Re-introducing patients as diagnostic agents. Int. J. Med. Inf. 80, e112–e126. doi:10.1016/j.ijmedinf.2010.09.010

Anderson, E.S., Winett, R.A., Wojcik, J.R., 2007. Self-regulation, self-efficacy, outcome expectations, and social support: Social cognitive theory and nutrition behavior. Ann. Behav. Med. 34, 304–312. doi:10.1007/BF02874555

© Springer Fachmedien Wiesbaden GmbH, part of Springer Nature 2018
D. D. Vaziri, *Facilitating Daily Life Integration of Technologies for Active and Healthy Aging*, Informationsmanagement in Theorie und Praxis,
https://doi.org/10.1007/978-3-658-22875-0

Anderson, E.S., Wojcik, J.R., Winett, R.A., Williams, D.M., 2006. Social-cognitive determinants of physical activity: The influence of social support, self-efficacy, outcome expectations, and self-regulation among participants in a church-based health promotion study. Health Psychol. 25, 510–520. doi:10.1037/0278-6133.25.4.510

Anderson-Hanley, C., Arciero, P.J., Brickman, A.M., Nimon, J.P., Okuma, N., Westen, S.C., Merz, M.E., Pence, B.D., Woods, J.A., Kramer, A.F., Zimmerman, E.A., 2012. Exergaming and Older Adult Cognition. Am. J. Prev. Med. 42, 109–119. doi:10.1016/j.amepre.2011.10.016

Andrews, G., Williams, A.D., 2014. INTERNET PSYCHOTHERAPY AND THE FUTURE OF PERSONALIZED TREATMENT: Commentary: Internet Psychotherapy. Depress. Anxiety 31, 912–915. doi:10.1002/da.22302

Arnhold, M., Quade, M., Kirch, W., 2014. Mobile Applications for Diabetics: A Systematic Review and Expert-Based Usability Evaluation Considering the Special Requirements of Diabetes Patients Age 50 Years or Older. J. Med. Internet Res. 16, e104. doi:10.2196/jmir.2968

Awad, M., Ferguson, S., Craig, C., 2014. Designing games for older adults: an affordance based approach. IEEE, pp. 1–7. doi:10.1109/SeGAH.2014.7067103

Bailey, B.P., Konstan, J.A., Carlis, J.V., 2001. The effects of interruptions on task performance, annoyance, and anxiety in the user interface, in: Proceedings of INTERACT. Citeseer, pp. 593–601.

Bainbridge, E., Bevans, S., Keeley, B., Oriel, K., 2011. The Effects of the Nintendo Wii Fit on Community-Dwelling Older Adults with Perceived Balance Deficits: A Pilot Study. Phys. Occup. Ther. Geriatr. 29, 126–135. doi:10.3109/02703181.2011.569053

Ballegaard, S.A., Hansen, T.R., Kyng, M., 2008. Healthcare in everyday life: designing healthcare services for daily life, in: Proceedings of the SIGCHI Conference on Human Factors in Computing Systems. Presented at the CHI '08, ACM Press, pp. 1807–1816. doi:10.1145/1357054.1357336

Bandura, A., 2004. Health Promotion by Social Cognitive Means. Health Educ. Behav. 31, 143–164. doi:10.1177/1090198104263660

Bandura, A., 1993. Perceived Self-Efficacy in Cognitive Development and Functioning. Educ. Psychol. 28, 117–148. doi:10.1207/s15326985ep2802_3

Bandura, A., 1986. Social foundations of thought and action: A social cognitive theory, Prentice-Hall series in social learning theory. Prentice-Hall, Inc, Englewood Cliffs, NJ, US.

Bandura, A., 1977. Self-efficacy: toward a unifying theory of behavioral change. Psychol. Rev. 84, 191.

Bangor, A., Kortum, P.T., Miller, J.T., 2008. An empirical evaluation of the system usability scale. Intl J. Human–Computer Interact. 24, 574–594.

Barnett, J., Harricharan, M., Fletcher, D., Gilchrist, B., Coughlan, J., 2015. myPace: an integrative health platform for supporting weight loss and maintenance behaviors. IEEE J. Biomed. Health Inform. 19, 109–116. doi:10.1109/JBHI.2014.2366832

Barnum, C.M., 2011. Usability testing essentials: ready, set-- test. Morgan Kaufmann Publishers, Burlington, MA.

Barry, A., Mcgwire, S., Porter, K., 2015. Global AgeWatch Index 2015 - Insight report.

Bateni, H., 2012. Changes in balance in older adults based on use of physical therapy vs the Wii Fit gaming system: a preliminary study. Physiotherapy 98, 211–216. doi:10.1016/j.physio.2011.02.004

Bell, C.S., Fain, E., Daub, J., Warren, S.H., Howell, S.H., Southard, K.S., Sellers, C., Shadoin, H., 2011. Effects of Nintendo Wii on Quality of Life, Social Relationships, and Confidence to Prevent Falls. Phys. Occup. Ther. Geriatr. 29, 213–221. doi:10.3109/02703181.2011.559307

Berkovsky, S., Coombe, M., Freyne, J., Bhandari, D., Baghaei, N., 2010. Physical activity motivating games: virtual rewards for real activity, in: Proceedings of the SIGCHI Conference on Human Factors in Computing Systems. Presented at the CHI '10, ACM Press, pp. 243–252. doi:10.1145/1753326.1753362

Berkovsky, S., Freyne, J., Coombe, M., 2012. Physical Activity Motivating Games: Be Active and Get Your Own Reward. ACM Trans. Comput.-Hum. Interact. 19, 1–41. doi:10.1145/2395131.2395139

Berwick, D.M., 2003. Disseminating innovations in health care. JAMA 289, 1969–1975. doi:10.1001/jama.289.15.1969

Betker, A.L., Szturm, T., Moussavi, Z.K., Nett, C., 2006. Video Game–Based Exercises for Balance Rehabilitation: A Single-Subject Design. Arch. Phys. Med. Rehabil. 87, 1141–1149. doi:10.1016/j.apmr.2006.04.010

Billipp, S.H., 2001. The Psychosocial Impact of Interactive Computer Use Within a Vulnerable Elderly Population: A Report on a Randomized Prospective Trial in a Home Health Care Setting. Public Health Nurs. 18, 138–145. doi:10.1046/j.1525-1446.2001.00138.x

Bisafar, F.I., Parker, A.G., 2016. Confidence & Control: Examining Adolescent Preferences for Technologies That Promote Wellness, in: Proceedings of the 19th ACM Conference on Computer-Supported Cooperative Work & Social Computing, CSCW '16. ACM, New York, NY, USA, pp. 160–171. doi:10.1145/2818048.2820028

Bjerknes, G., Bratteteig, T., 1995. User Participation and Democracy: A Discussion of Scandinavian Research on Systems Development. Scand J Inf Syst 7, 73–98.

Bleakley, C.M., Charles, D., Porter-Armstrong, A., McNeill, M.D.J., McDonough, S.M., McCormack, B., 2015. Gaming for Health: A Systematic Review of the Physical and Cognitive Effects of Interactive Computer Games in Older Adults. J. Appl. Gerontol. 34, NP166-NP189. doi:10.1177/0733464812470747

Bodker, K., Kensing, F., Simonsen, J., 2010. Participatory Design in Information Systems Development, in: Isomäki, H., Pekkola, S. (Eds.), Reframing Humans in Information Systems Development. Springer London, London, pp. 115–134. doi:10.1007/978-1-84996-347-3_7

Bodker, K., Kensing, F., Simonsen, J., 2004. Participatory It Design: Designing for Business and Workplace Realities. MIT Press, Cambridge, MA, USA.

Bomberger, S.A., 2010. The effects of Nintendo Wii Fit on balance of elderly adults.

Borsci, S., Federici, S., Lauriola, M., 2009. On the dimensionality of the System Usability Scale: a test of alternative measurement models. Cogn. Process. 10, 193–197. doi:10.1007/s10339-009-0268-9

Borson, S., Scanlan, J., Brush, M., Vitaliano, P., Dokmak, A., 2000. The Mini-Cog: a cognitive"vital signs" measure for dementia screening in multi-lingual elderly. Int. J. Geriatr. Psychiatry 15, 1021–1027.

Botella, C., Mira, A., Garcia-Palacios, A., Quero, S., Navarro, M.V., Riera López Del Amo, A., Molinari, G., Castilla, D., Moragrega, I., Soler, C., Alcañiz, M., Baños, R.M., 2012. Smiling is fun: a Coping with Stress and Emotion Regulation Program. Stud. Health Technol. Inform. 181, 123–127.

Bradley, N., Poppen, W., 2003. Assistive technology, computers and Internet may decrease sense of isolation for homebound elderly and disabled persons. Technol. Disabil. 15, 19–25.

Brannen, J., 2005. Mixing Methods: The Entry of Qualitative and Quantitative Approaches into the Research Process. Int. J. Soc. Res. Methodol. 8, 173–184. doi:10.1080/13645570500154642

Braun, M.T., 2013. Obstacles to social networking website use among older adults. Comput. Hum. Behav. 29, 673–680. doi:10.1016/j.chb.2012.12.004

Braun, V., Clarke, V., 2006. Using thematic analysis in psychology. Qual. Res. Psychol. 3, 77–101. doi:10.1191/1478088706qp063oa

Brauner, P., Calero Valdez, A., Schroeder, U., Ziefle, M., 2013. Increase Physical Fitness and Create Health Awareness through Exergames and Gamification, in: Holzinger, A., Ziefle, M., Hitz, M., Debevc, M. (Eds.), Human Factors in Computing and Informatics. Springer Berlin Heidelberg, Berlin, Heidelberg, pp. 349–362.

Bravata, D.M., Smith-Spangler, C., Sundaram, V., Gienger, A.L., Lin, N., Lewis, R., Stave, C.D., Olkin, I., Sirard, J.R., 2007. Using pedometers to increase physical activity and improve health: a systematic review. JAMA 298, 2296–2304. doi:10.1001/jama.298.19.2296

Brodaty, H., Thomson, C., Thompson, C., Fine, M., 2005. Why caregivers of people with dementia and memory loss don't use services. Int. J. Geriatr. Psychiatry 20, 537–546. doi:10.1002/gps.1322

Brodie, M., Lord, S., Coppens, M., Annegarn, J., Delbaere, K., 2015. Eight weeks remote monitoring using a freely worn device reveals unstable gait patterns in older fallers. IEEE Trans. Biomed. Eng. 1–1. doi:10.1109/TBME.2015.2433935

Broekhuizen, K., Kroeze, W., van Poppel, M.N., Oenema, A., Brug, J., 2012. A Systematic Review of Randomized Controlled Trials on the Effectiveness of Computer-Tailored Physical Activity and Dietary Behavior Promotion Programs: an Update. Ann. Behav. Med. 44, 259–286. doi:10.1007/s12160-012-9384-3

Brooke, J., 1996. SUS: a quick and dirty usability scale. PW Jordan PW Thomas B Weerdmeester BA McClelland IL Eds Usability Eval. Ind. Taylor Francis Lond.

Brox, E., Luque, L.F., Evertsen, G.J., Hernández, J.E.G., 2011. Exergames for elderly: Social exergames to persuade seniors to increase physical activity, in: Pervasive Computing Technologies for Healthcare (PervasiveHealth), 2011 5th International Conference on. IEEE, pp. 546–549.

Bryman, A., 2012. Social Research Methods. OUP Oxford.

Budweg, S., Lewkowicz, M., Müller, C., Schering, S., 2012. Fostering Social Interaction in AAL: Methodological reflections on the coupling of real household Living Lab and SmartHome approaches. -Com 11, 30–35. doi:10.1524/icom.2012.0035

Buhler, C., 1935. The curve of life as studied in biographies. J. Appl. Psychol. 19, 405–409. doi:10.1037/h0054778

Buhler, C.M., Massarik, F., Bugental, J.F.T., 1968. The course of human life: a study of goals in the humanistic perspective. Springer Pub. Co.

Campbell, A.J., Robertson, M.C., Gardner, M.M., Norton, R.N., Tilyard, M.W., Buchner, D.M., 1997. Randomised controlled trial of a general practice programme of home based exercise to prevent falls in elderly women. BMJ 315, 1065–1069. doi:10.1136/bmj.315.7115.1065

Campbell, T., Ngo, B., Fogarty, J., 2008. Game design principles in everyday fitness applications, in: Proceedings of the 2008 ACM Conference on Computer Supported Cooperative Work. Presented at the CSCW '08, ACM Press, pp. 249–252. doi:10.1145/1460563.1460603

Cantwell, D., Broin, D.O., Palmer, R., Doyle, G., 2012. Motivating Elderly People to Exercise Using a Social Collaborative Exergame with Adaptive Difficulty., in: Proceeding of the 6th European Conference on Games Based Learning–ECGBL.

Carmichael, A., Rice, M., MacMillan, F., Kirk, A., 2010. Investigating a DTV-based physical activity application to facilitate wellbeing in older adults, in: Proceedings of the 24th BCS Interaction Specialist Group Conference. British Computer Society, pp. 278–288.

Carmien, S., Manzanares, A.G., 2014. Elders Using Smartphones – A Set of Research Based Heuristic Guidelines for Designers, in: Stephanidis, C., Antona, M. (Eds.), Universal Access in Human-Computer Interaction. Universal Access to Information and Knowledge. Springer International Publishing, Cham, pp. 26–37. doi:10.1007/978-3-319-07440-5_3

Carroll, J.M., Rosson, M.B., 2013. Wild at Home: The Neighborhood as a Living Laboratory for HCI. ACM Trans. Comput.-Hum. Interact. 20, 1–28. doi:10.1145/2491500.2491504

Centers for Disease Control and Prevention, 2013. The state of aging and health in America 2013. Atlanta GA Cent. Dis. Control Prev. US Dep. Health Hum. Serv.

Chao, D., Foy, C.G., Farmer, D., 2000. Exercise Adherence among Older Adults. Control. Clin. Trials 21, S212–S217. doi:10.1016/S0197-2456(00)00081-7

Charness, N., Schaie, K.W., 2003. Impact of Technology on Successful Aging, Springer Series on the Societal Impact on Aging. Springer Publishing Company.

Chaudhry, B., Duarte, M., Chawla, N.V., Dasgupta, D., 2016. Developing Health Technologies for Older Adults: Methodological and Ethical Considerations, in: Proceedings of the 10th EAI International Conference on Pervasive Computing Technologies for Healthcare, PervasiveHealth '16. ICST (Institute for Computer Sciences, Social-Informatics and Telecommunications Engineering), ICST, Brussels, Belgium, Belgium, pp. 330–332.

Chen, K., Chan, A.H.S., 2011. A review of technology acceptance by older adults. Gerontechnology 10. doi:10.4017/gt.2011.10.01.006.00

Chen, Y., Ngo, V., Park, S.Y., 2013. Caring for caregivers: designing for integrality, in: Proceedings of the 2013 Conference on Computer Supported Cooperative

Work. Presented at the CSCW '13, ACM Press, pp. 91–102. doi:10.1145/2441776.2441789

Chiauzzi, E., Rodarte, C., DasMahapatra, P., 2015. Patient-centered activity monitoring in the self-management of chronic health conditions. BMC Med. 13. doi:10.1186/s12916-015-0319-2

Choe, E.K., Lee, N.B., Lee, B., Pratt, W., Kientz, J.A., 2014. Understanding quantified-selfers' practices in collecting and exploring personal data, in: Proceedings of the SIGCHI Conference on Human Factors in Computing Systems Pages 1143-1152. Presented at the CHI '14, ACM Press, pp. 1143–1152. doi:10.1145/2556288.2557372

Cialdini, R.B., 2001. Influence: Science and Practice. Allyn and Bacon.

Cila, N., Jansen, G., Groen, M., Meys, W., den Broeder, L., Kröse, B., 2016. Look! A Healthy Neighborhood: Means to Motivate Participants in Using an App for Monitoring Community Health, in: Proceedings of the 2016 CHI Conference Extended Abstracts on Human Factors in Computing Systems, CHI EA '16. ACM, New York, NY, USA, pp. 889–898. doi:10.1145/2851581.2851591

Cimperman, M., Brenčič, M.M., Trkman, P., Stanonik, M. de L., 2013. Older Adults' Perceptions of Home Telehealth Services. Telemed. E-Health 19, 786–790. doi:10.1089/tmj.2012.0272

Clement, A., Van den Besselaar, P., 1993. A retrospective look at PD projects. Commun. ACM - Spec. Issue Particip. Des. 36, 29–37. doi:10.1145/153571.163264

Coddington, D.C. (Ed.), 2000. Beyond managed care: how consumers and technology are changing the future of health care, 1st ed. ed, Jossey-Bass health care series. Jossey-Bass, San Francisco.

Conn, V.S., Hafdahl, A.R., Brown, L.M., 2009. Meta-analysis of Quality-of-Life Outcomes From Physical Activity Interventions: Nurs. Res. 58, 175–183. doi:10.1097/NNR.0b013e318199b53a

Consolvo, S., Everitt, K., Smith, I., Landay, J.A., 2006. Design requirements for technologies that encourage physical activity, in: Proceedings of the SIGCHI Conference on Human Factors in Computing Systems. Presented at the CHI '06, ACM Press, pp. 457–466. doi:10.1145/1124772.1124840

Consolvo, S., Libby, R., Smith, I., Landay, J.A., McDonald, D.W., Toscos, T., Chen, M.Y., Froehlich, J., Harrison, B., Klasnja, P., LaMarca, A., LeGrand, L., 2008. Activity sensing in the wild: a field trial of ubifit garden, in: Proceedings of the SIGCHI Conference on Human Factors in Computing Systems. Presented at the CHI '08, ACM Press, pp. 1797–1806. doi:10.1145/1357054.1357335

Cook, D.J., Manning, D.M., Holland, D.E., Prinsen, S.K., Rudzik, S.D., Roger, V.L., Deschamps, C., 2013. Patient Engagement and Reported Outcomes in Surgical Recovery: Effectiveness of an e-Health Platform. J. Am. Coll. Surg. 217, 648–655. doi:10.1016/j.jamcollsurg.2013.05.003

Corbin, J., Strauss, A., 2014. Basics of qualitative research: Techniques and procedures for developing grounded theory. Sage Publications.

Corti, K., 2006. Games-based Learning; a serious business application. Inf. PixelLearning 34, 1–20.

Cotten, S.R., Anderson, W.A., McCullough, B.M., 2013. Impact of Internet Use on Loneliness and Contact with Others Among Older Adults: Cross-Sectional Analysis. J. Med. Internet Res. 15, e39. doi:10.2196/jmir.2306

Coughlin, J., D'Ambrosio, L.A., Reimer, B., Pratt, M.R., 2007. Older adult perceptions of smart home technologies: implications for research, policy & market innovations in healthcare. Conf. Proc. Annu. Int. Conf. IEEE Eng. Med. Biol. Soc. IEEE Eng. Med. Biol. Soc. Annu. Conf. 2007, 1810–1815. doi:10.1109/IEMBS.2007.4352665

Czaja, S.J., Charness, N., Fisk, A.D., Hertzog, C., Nair, S.N., Rogers, W.A., Sharit, J., 2006. Factors predicting the use of technology: Findings from the center for research and education on aging and technology enhancement (create). Psychol. Aging 21, 333–352. doi:10.1037/0882-7974.21.2.333

Daniel, K., 2012. Wii-Hab for Pre-Frail Older Adults. Rehabil. Nurs. 37, 195–201. doi:10.1002/rnj.25

Davis, F.D., 1989. Perceived Usefulness, Perceived Ease of Use, and User Acceptance of Information Technology. MIS Q. 13, 319. doi:10.2307/249008

Day, L., Fildes, B., Gordon, I., Fitzharris, M., Flamer, H., Lord, S., 2002. Randomised factorial trial of falls prevention among older people living in their own homes. BMJ 325, 128.

de Bruin, E.D., Hartmann, A., Uebelhart, D., Murer, K., Zijlstra, W., 2008. Wearable systems for monitoring mobility-related activities in older people: a systematic review. Clin. Rehabil. 22, 878–895. doi:10.1177/0269215508090675

Dent, E., Kowal, P., Hoogendijk, E.O., 2016. Frailty measurement in research and clinical practice: A review. Eur. J. Intern. Med. 31, 3–10. doi:10.1016/j.ejim.2016.03.007

Denzin, N.K., 1970. The Research Act in Sociology: A Theoretical Introduction to Sociological Methods, Methodological perspectives. Butterworths.

Di Pasquale, D., Padula, M., Scala, P.L., Biocca, L., Paraciani, N., 2013. Advancements in ICT for healthcare and wellbeing: towards Horizon 2020, in: Proceedings of the Fifth International Conference on Management of Emergent Digital Eco-Systems. Presented at the MEDES '13, ACM Press, pp. 353–358. doi:10.1145/2536146.2536201

Dishman, E., 2004. Inventing wellness systems for aging in place. Computer 37, 34–41. doi:10.1109/MC.2004.1297237

Dittrich, Y., Eriksén, S., Hansson, C., 2002. PD in the Wild; Evolving Practices of Design in Use, in: Proceedings of the Participatory Design Conference. Presented at the PDC 02, pp. 124–134.

Dorst, K., 2006. Design Problems and Design Paradoxes. Des. Issues 22, 4–17. doi:10.1162/desi.2006.22.3.4

Doyle, J., Bailey, C., Dromey, B., Scanaill, C.N., 2010a. BASE - An interactive technology solution to deliver balance and strength exercises to older adults, in: Pervasive Computing Technologies for Healthcare (PervasiveHealth). Presented at the 4th International Conference on-NO PERMISSIONS, IEEE. doi:10.4108/ICST.PERVASIVEHEALTH2010.8881

Doyle, J., Skrba, Z., McDonnell, R., Arent, B., 2010b. Designing a Touch Screen Communication Device to Support Social Interaction Amongst Older Adults, in: Proceedings of the 24th BCS Interaction Specialist Group Conference, BCS '10. British Computer Society, Swinton, UK, UK, pp. 177–185.

Doyle, J., Walsh, L., Sassu, A., McDonagh, T., 2014. Designing a Wellness Self-Management Tool for Older Adults – Results from a Field Trial of YourWellness, in: Proceedings of the 8th International Conference on Pervasive Computing Technologies for Healthcare. Presented at the PervasiveHealth '14, ICST, pp. 134–141. doi:10.4108/icst.pervasivehealth.2014.254950

Drabble, S.J., O'Cathain, A., Thomas, K.J., Rudolph, A., Hewison, J., 2014. Describing qualitative research undertaken with randomised controlled trials in grant proposals: a documentary analysis. BMC Med. Res. Methodol. 14, 24. doi:10.1186/1471-2288-14-24

Eberst, R.M., 1984. Defining Health: A Multidimensional Model. J. Sch. Health 54, 99–104. doi:10.1111/j.1746-1561.1984.tb08780.x

Ebert, H.H., Heimermann, M., 2004. Einfach - fuer den Menschen! Grundaspekte Usability.

Ehn, P., 2008. Participation in Design Things, in: Proceedings of the Tenth Anniversary Conference on Participatory Design 2008. Presented at the CSCW, Indiana University, Indianapolis, IN, USA, pp. 92–101.

Ehn, P., 1990. Work-oriented design of computer artifacts. Erlbaum Associates Inc.

Eisma, R., Dickinson, A., Goodman, J., Mival, O., Syme, A., Tiwari, L., 2003. Mutual inspiration in the development of new technology for older people. Presented at the Proceedings of Include, Citeseer, pp. 252–259.

El-Khoury, F., Cassou, B., Charles, M.-A., Dargent-Molina, P., 2015. The effect of fall prevention exercise programmes on fall induced injuries in community dwelling older adults: Br. J. Sports Med. 49, 1348–1348. doi:10.1136/bmj.f6234

Erikson, E.H., 1995. Childhood and Society, Pelican Book. Vintage.

European Commission, 2013. European Commission - PRESS RELEASES - Press release - Active & Healthy Ageing: EU cities and regions get star ratings to recognise excellent projects [WWW Document]. URL http://europa.eu/rapid/press-release_IP-13-633_en.htm (accessed 6.28.17).

Evenson, K.R., Goto, M.M., Furberg, R.D., 2015. Systematic review of the validity and reliability of consumer-wearable activity trackers. Int. J. Behav. Nutr. Phys. Act. 12. doi:10.1186/s12966-015-0314-1

Fanning, J., Mullen, S.P., McAuley, E., 2012. Increasing Physical Activity With Mobile Devices: A Meta-Analysis. J. Med. Internet Res. 14, e161. doi:10.2196/jmir.2171

Fausset, C.B., Mitzner, T.L., Price, C.E., Jones, B.D., Fain, B.W., Rogers, W.A., 2013. Older Adults' Use of and Attitudes toward Activity Monitoring Technologies. Proc. Hum. Factors Ergon. Soc. Annu. Meet. 57, 1683–1687. doi:10.1177/1541931213571374

Fazeli, P.L., Ross, L.A., Vance, D.E., Ball, K., 2013. The Relationship Between Computer Experience and Computerized Cognitive Test Performance Among Older Adults. J. Gerontol. B. Psychol. Sci. Soc. Sci. 68, 337–346. doi:10.1093/geronb/gbs071

Ferrucci, L., Baldasseroni, S., Bandinelli, S., de Alfieri, W., Cartei, A., Calvani, D., Baldini, A., Masotti, G., Marchionni, N., 2000. Disease severity and health-related quality of life across different chronic conditions. J. Am. Geriatr. Soc. 48, 1490–1495.

Finkelstein, E.A., Haaland, B.A., Bilger, M., Sahasranaman, A., Sloan, R.A., Nang, E.E.K., Evenson, K.R., 2016. Effectiveness of activity trackers with and without incentives to increase physical activity (TRIPPA): a randomised controlled trial. Lancet Diabetes Endocrinol. 4, 983–995. doi:10.1016/S2213-8587(16)30284-4

Fischer, S.H., David, D., Crotty, B.H., Dierks, M., Safran, C., 2014. Acceptance and use of health information technology by community-dwelling elders. Int. J. Med. Inf. 83, 624–635. doi:10.1016/j.ijmedinf.2014.06.005

Fitzpatrick, G., Ellingsen, G., 2013. A Review of 25 Years of CSCW Research in Healthcare: Contributions, Challenges and Future Agendas. Comput. Support. Coop. Work CSCW 22, 609–665. doi:10.1007/s10606-012-9168-0

Fogg, B.J., 2009. A behavior model for persuasive design, in: Proceedings of the 4th International Conference on Persuasive Technology. ACM, p. 40.

Fogg, B.J., 2007. Mobile Persuasion: 20 Perspectives on the Future of Behavior Change. Stanford Captology Media.

Fogg, B.J., 2002. Persuasive technology: using computers to change what we think and do. Ubiquity 2002, 2. doi:10.1145/764008.763957

Freiberger, E., Menz, H.B., Abu-Omar, K., Rutten, A., 2007. Preventing falls in physically active community-dwelling older people: a comparison of two intervention techniques. Gerontology 53, 298–305. doi:10.1159/000103256

Fritz, T., Huang, E.M., Murphy, G.C., Zimmermann, T., 2014. Persuasive Technology in the Real World: A Study of Long-term Use of Activity Sensing Devices for Fitness, in: Proceedings of the SIGCHI Conference on Human Factors in Computing Systems, CHI '14. ACM, New York, NY, USA, pp. 487–496. doi:10.1145/2556288.2557383

Fuchs, R., 1997. Psychologie und körperliche Bewegung: Grundlagen für theoriegeleitete Interventionen, Reihe Gesundheitspsychologie. Hogrefe, Verlag für Psychologie.

Fuchs, R., 1994. Konsequenzerwartungen als Determinante des Sport- und Bewegungsverhaltens. Z. Für Gesundheitspsychologie 2, S. 269-291.

Fuchs, R., Schwarzer, R., 1994. Selbstwirksamkeit zur sportlichen Aktivität: Reliabilität und Validität eines neuen Meßinstrumen. Z. Für Differ. Diagn. Psychol. 15, 141–154.

Gagnon, M.-P., Légaré, F., Labrecque, M., Frémont, P., Pluye, P., Gagnon, J., Gravel, K., 2006. Interventions for promoting information and communication technologies adoption in healthcare professionals, in: The Cochrane Collaboration (Ed.), Cochrane Database of Systematic Reviews. John Wiley & Sons, Ltd, Chichester, UK. doi:10.1002/14651858.CD006093

Gaßner, K.S., 2010. ICT enabled independent living for the elderly a status-quo analysis on products and the research landscape in the field of ambient assisted living (AAL) in EU-27. IIT, Berlin.

Geraedts, H.A., Zijlstra, W., Zhang, W., Bulstra, S., Stevens, M., 2014. Adherence to and effectiveness of an individually tailored home-based exercise program for frail older adults, driven by mobility monitoring: design of a prospective cohort study. BMC Public Health 14, 570. doi:10.1186/1471-2458-14-570

Gerling, K., Livingston, I., Nacke, L., Mandryk, R., 2012. Full-body Motion-based Game Interaction for Older Adults, in: Proceedings of the SIGCHI Conference on Human Factors in Computing Systems, CHI '12. ACM, New York, NY, USA, pp. 1873–1882. doi:10.1145/2207676.2208324

Gerling, K.M., Mandryk, R.L., Linehan, C., 2015. Long-Term Use of Motion-Based Video Games in Care Home Settings, in: Proceedings of the 33rd Annual ACM Conference on Human Factors in Computing Systems, CHI '15. ACM, New York, NY, USA, pp. 1573–1582. doi:10.1145/2702123.2702125

Gerling, K.M., Masuch, M., 2011. Exploring the potential of gamification among frail elderly persons, in: Proceedings of the CHI 2011 Workshop Gamification: Using Game Design Elements in Non-Game Contexts.

Giacomini, M., Baylis, F., Robert, J., 2007. Banking on it: Public policy and the ethics of stem cell research and development. Soc. Sci. Med. 65, 1490–1500. doi:10.1016/j.socscimed.2007.05.021

Gill, T.M., Pahor, M., Guralnik, J.M., McDermott, M.M., King, A.C., Buford, T.W., Strotmeyer, E.S., Nelson, M.E., Sink, K.M., Demons, J.L., Kashaf, S.S., Walkup, M.P., Miller, M.E., 2016. Effect of structured physical activity on prevention of serious fall injuries in adults aged 70-89: randomized clinical trial (LIFE Study). BMJ i245. doi:10.1136/bmj.i245

Gillespie, L.D., Robertson, M.C., Gillespie, W.J., Sherrington, C., Gates, S., Clemson, L.M., Lamb, S.E., 2012. Interventions for preventing falls in older people living in the community, in: The Cochrane Collaboration, Gillespie, L.D. (Eds.), Cochrane Database of Systematic Reviews. John Wiley & Sons, Ltd, Chichester, UK.

Göbel, S., Hardy, S., Wendel, V., Mehm, F., Steinmetz, R., 2010. Serious games for health: personalized exergames, in: Proceedings of the 18th ACM International Conference on Multimedia. Presented at the MM '10, ACM Press, p. 1663. doi:10.1145/1873951.1874316

Goldberg, J.H., King, A.C., 2007. Physical Activity and Weight Management Across the Lifespan. Annu. Rev. Public Health 28, 145–170. doi:10.1146/annurev.publhealth.28.021406.144105

Goldstein, C.M., Gathright, E.C., Dolansky, M.A., Gunstad, J., Sterns, A., Redle, J.D., Josephson, R., Hughes, J.W., 2014. Randomized controlled feasibility trial of two telemedicine medication reminder systems for older adults with heart failure. J. Telemed. Telecare 20, 293–299. doi:10.1177/1357633X14541039

Google, 2017a. Color - Style [WWW Document]. Mater. Des. Guidel. URL https://material.io/guidelines/style/color.html#color-color-palette (accessed 5.23.17).

Google, 2017b. Buttons - Components [WWW Document]. Mater. Des. Guidel. URL https://material.io/guidelines/components/buttons.html# (accessed 5.23.17).

Graham, D.J., Hipp, J.A., Marshall, S., Kerr, J., 2014. Emerging Technologies to Promote and Evaluate Physical Activity:, Frontiers Research Topics. Frontiers E-books.

Green, C.A., Duan, N., Gibbons, R.D., Hoagwood, K.E., Palinkas, L.A., Wisdom, J.P., 2015. Approaches to Mixed Methods Dissemination and Implementation Research: Methods, Strengths, Caveats, and Opportunities. Adm. Policy Ment. Health Ment. Health Serv. Res. 42, 508–523. doi:10.1007/s10488-014-0552-6

Grindrod, K.A., Li, M., Gates, A., 2014. Evaluating User Perceptions of Mobile Medication Management Applications With Older Adults: A Usability Study. JMIR Mhealth Uhealth 2, e11. doi:10.2196/mhealth.3048

Grönvall, E., Kyng, M., 2013. On participatory design of home-based healthcare. Cogn. Technol. Work 15, 389–401. doi:10.1007/s10111-012-0226-7

Gschwind, Y.J., Eichberg, S., Ejupi, A., de Rosario, H., Kroll, M., Marston, H.R., Drobics, M., Annegarn, J., Wieching, R., Lord, S.R., Aal, K., Vaziri, D., Woodbury, A., Fink, D., Delbaere, K., 2015. ICT-based system to predict and prevent falls (iStoppFalls): results from an international multicenter randomized controlled trial. Eur. Rev. Aging Phys. Act. 12. doi:10.1186/s11556-015-0155-6

Gschwind, Y.J., Eichberg, S., Marston, H.R., Ejupi, A., de Rosario, H., Kroll, M., Drobics, M., Annegarn, J., Wieching, R., Lord, S.R., Aal, K., Delbaere, K., 2014. ICT-based system to predict and prevent falls (iStoppFalls): study protocol for an international multicenter randomized controlled trial. BMC Geriatr. 14, 91. doi:10.1186/1471-2318-14-91

Gualtieri, L., Rosenbluth, S., Phillips, J., 2016. Can a Free Wearable Activity Tracker Change Behavior? The Impact of Trackers on Adults in a Physician-Led Wellness Group. JMIR Res. Protoc. 5, e237. doi:10.2196/resprot.6534

Gudur, R.R., Blackler, A., Popovic, V., Mahar, D., 2013. Ageing, Technology Anxiety and Intuitive Use of Complex Interfaces, in: Kotzé, P., Marsden, G., Lindgaard, G., Wesson, J., Winckler, M. (Eds.), Human-Computer Interaction – INTERACT 2013. Springer Berlin Heidelberg, Berlin, Heidelberg, pp. 564–581. doi:10.1007/978-3-642-40477-1_36

Guedes, D.P., Hatmann, A.C., Martini, F.A.N., Borges, M.B., Bernardelli, R., 2012. Quality of Life and Physical Activity in a Sample of Brazilian Older Adults. J. Aging Health 24, 212–226. doi:10.1177/0898264311410693

Haberer, J.E., Robbins, G.K., Ybarra, M., Monk, A., Ragland, K., Weiser, S.D., Johnson, M.O., Bangsberg, D.R., 2012. Real-Time Electronic Adherence Monitoring is Feasible, Comparable to Unannounced Pill Counts, and Acceptable. AIDS Behav. 16, 375–382. doi:10.1007/s10461-011-9933-y

Halko, S., Kientz, J.A., 2010. Personality and Persuasive Technology: An Exploratory Study on Health-Promoting Mobile Applications, in: Ploug, T., Hasle, P., Oinas-Kukkonen, H. (Eds.), Persuasive Technology. Springer Berlin Heidelberg, Berlin, Heidelberg, pp. 150–161. doi:10.1007/978-3-642-13226-1_16

Hall, K.S., Crowley, G.M., Bosworth, H.B., Howard, T.A., Morey, M.C., 2010. Individual progress toward self-selected goals among older adults enrolled in a physical activity counseling intervention. J. Aging Phys. Act. 18, 439–450.

Haluza, D., Jungwirth, D., 2015. ICT and the future of health care: aspects of health promotion. Int. J. Med. Inf. 84, 48–57. doi:10.1016/j.ijmedinf.2014.09.005

Hammersley, M., 2000. Varieties of social research: A typology. Int. J. Soc. Res. Methodol. 3, 221–229. doi:10.1080/13645570050083706

Hammersley, M., 1996. The relationship between qualitative and quantitative research: paradigm loyalty versus methodological eclecticism, in: Richardson, J.T.E. (Ed.), Handbook of Qualitative Research Methods for Psychology and the Social Sciences. British Psychological Society, Leicester.

Hänsel, K., Wilde, N., Haddadi, H., Alomainy, A., 2015. Challenges with Current Wearable Technology in Monitoring Health Data and Providing Positive Behavioural Support. ICST. doi:10.4108/eai.14-10-2015.2261601

Hartswood, M., Procter, R., Slack, R., Voß, A., Büscher, M., Rouncefield, M., Rouchy, P., 2008. Co-Realization: Toward a Principled Synthesis of Ethnomethodology and Participatory Design, in: Resources, Co-Evolution and Artifacts. Springer London, London, pp. 59–94.

Heaney, C. A., & Israel, B.A., 2008. Social networks and social support. Health behavior and health education: Theory, research, and practice, 4,.

Heart, T., Kalderon, E., 2013. Older adults: Are they ready to adopt health-related ICT? Int. J. Med. Inf. 82, e209–e231. doi:10.1016/j.ijmedinf.2011.03.002

Hernández-Encuentra, E., Pousada, M., Gómez-Zúñiga, B., 2009. ICT and Older People: Beyond Usability. Educ. Gerontol. 35, 226–245. doi:10.1080/03601270802466934

Herzberg, F., 1966. Work and the nature of man. World Pub. Co.

Herzberg, F., Mausner, B., Snyderman, B.B., 2011. The Motivation to Work, Organization and Business. Transaction Publishers.

Huber, M., Knottnerus, J.A., Green, L., Horst, H. v. d., Jadad, A.R., Kromhout, D., Leonard, B., Lorig, K., Loureiro, M.I., Meer, J.W.M. v. d., Schnabel, P., Smith, R., Weel, C. v., Smid, H., 2011. How should we define health? BMJ 343, d4163–d4163. doi:10.1136/bmj.d4163

Ihde, D., 1990. Technology and the Lifeworld: From Garden to Earth, Midland book. Indiana University Press.

Ihde, D., Selinger, E., 2003. Chasing Technoscience: Matrix for Materiality, Indiana series in the philosophy of technology. Indiana University Press.

Ijsselsteijn, W., Nap, H.H., de Kort, Y., Poels, K., 2007. Digital game design for elderly users, in: Proceedings of the 2007 Conference on Future Play. Presented at the Future Play '07, ACM Press, pp. 17–22. doi:10.1145/1328202.1328206

Jacobs, M.L., Clawson, J., Mynatt, E.D., 2015. Comparing Health Information Sharing Preferences of Cancer Patients, Doctors, and Navigators, in: Proceedings of the 18th ACM Conference on Computer Supported Cooperative Work & Social Computing, CSCW '15. ACM, New York, NY, USA, pp. 808–818. doi:10.1145/2675133.2675252

Jacqueline K. Eastman, Rajesh Iyer, 2004. The elderly's uses and attitudes towards the Internet. J. Consum. Mark. 21, 208–220. doi:10.1108/07363760410534759

Jakobs, E.-M., Lehnen, K., Ziefle, M., 2008. Alter und Technik: Studie zu Technikkonzepten, Techniknutzung und Technikbewertung älterer Menschen. BoD–Books on Demand.

Janssen, S., Tange, H., Arends, R., 2013. A Preliminary Study on the Effectiveness of Exergame Nintendo "Wii Fit Plus" on the Balance of Nursing Home Residents. Games Health J. 2, 89–95. doi:10.1089/g4h.2012.0074

Jarman, H., 2014. Incentivizing health information exchange: collaborative governance, market failure, and the public interest, in: Proceedings of the 15th Annual International Conference on Digital Government Research. Presented at the dg.o '14, ACM Press, pp. 227–235. doi:10.1145/2612733.2612766

Jimison, H.B., Pavel, M., Hatt, W.J., Chan, M., Larimer, N., Yu, C.H., 2010. Delivering a multi-faceted cognitive health intervention to the home. Gerontechnology 9. doi:10.4017/gt.2010.09.02.297.00

Jorgensen, M.G., Laessoe, U., Hendriksen, C., Nielsen, O.B.F., Aagaard, P., 2013. Efficacy of Nintendo Wii Training on Mechanical Leg Muscle Function and Postural Balance in Community-Dwelling Older Adults: A Randomized Controlled Trial. J. Gerontol. A. Biol. Sci. Med. Sci. 68, 845–852. doi:10.1093/gerona/gls222

Jung, Y., Li, K.J., Janissa, N.S., Gladys, W.L.C., Lee, K.M., 2009. Games for a better life: effects of playing Wii games on the well-being of seniors in a long-term care facility, in: Proceedings of the Sixth Australasian Conference on Interactive Entertainment. Presented at the IE '09, ACM Press, pp. 1–6. doi:10.1145/1746050.1746055

Kahn, E.B., Ramsey, L.T., Brownson, R.C., Heath, G.W., Howze, E.H., Powell, K.E., Stone, E.J., Rajab, M.W., Corso, P., 2002. The effectiveness of interventions to increase physical activity. A systematic review. Am. J. Prev. Med. 22, 73–107.

Kannus, P., Sievänen, H., Palvanen, M., Järvinen, T., Parkkari, J., 2005. Prevention of falls and consequent injuries in elderly people. The Lancet 366, 1885–1893. doi:10.1016/S0140-6736(05)67604-0

Kari, T., Makkonen, M., Moilanen, P., Frank, L., 2012. The habits of playing and the reasons for not playing exergames: Gender differences in Finland. 25th Bled

EConference EDependability Reliab. Trust. EStructures EProcesses EOperations Eser. Future Res. Vol.-17-206 2012 Bled Slov. Pp 512-526 Ed. U Lechner Wigand Pucihar ISBN 978-961-232-256-4.

Keith, S., Whitney, G., 1998. Bridging the gap between young designers and older users in an inclusive society. Proc Good Bad Challenging User Future ICT.

Kendzierski, D., DeCarlo, K.J., 1991. Physical Activity Enjoyment Scale: Two Validation Studies. J. Sport Exerc. Psychol. 13.

Khvorostianov, N., Elias, N., Nimrod, G., 2012. "Without it I am nothing": The internet in the lives of older immigrants. New Media Soc. 14, 583–599. doi:10.1177/1461444811421599

Kiosses, D.N., Leon, A.C., Areán, P.A., 2011. Psychosocial Interventions for Late-life Major Depression: Evidence-Based Treatments, Predictors of Treatment Outcomes, and Moderators of Treatment Effects. Psychiatr. Clin. North Am. 34, 377–401. doi:10.1016/j.psc.2011.03.001

Klasnja, P., Pratt, W., 2012. Healthcare in the pocket: Mapping the space of mobile-phone health interventions. J. Biomed. Inform. 45, 184–198. doi:10.1016/j.jbi.2011.08.017

Knight, E., Stuckey, M.I., Petrella, R.J., 2014. Health Promotion Through Primary Care: Enhancing Self-Management With Activity Prescription and mHealth. Phys. Sportsmed. 42, 90–99. doi:10.3810/psm.2014.09.2080

Kothgassner, O.D., Felnhofer, A., Hauk, N., Kastenhofer, E., Gomm, J., Krysprin-Exner, I., 2013. Technology Usage Inventory.

Krämer, L., Fuchs, R., 2010. Barrieren und Barrierenmanagement im Prozess der Sportteilnahme: Zwei neue Messinstrumente. Z. Für Gesundheitspsychologie 18, 170–182. doi:10.1026/0943-8149/a000026

Lai, C.-H., Peng, C.-W., Chen, Y.-L., Huang, C.-P., Hsiao, Y.-L., Chen, S.-C., 2013. Effects of interactive video-game based system exercise on the balance of the elderly. Gait Posture 37, 511–515. doi:10.1016/j.gaitpost.2012.09.003

Lamoth Claudine, J.C., Caljouw Simone, R., Postema, K., 2011. Active Video Gaming to Improve Balance in the Elderly. Stud. Health Technol. Inform. 159–164. doi:10.3233/978-1-60750-766-6-159

Lane, N.D., Lin, M., Mohammod, M., Yang, X., Lu, H., Cardone, G., Ali, S., Doryab, A., Berke, E., Campbell, A.T., Choudhury, T., 2014. BeWell: Sensing Sleep, Physical Activities and Social Interactions to Promote Wellbeing. Mob. Netw. Appl. 19, 345–359. doi:10.1007/s11036-013-0484-5

Lansdowne, A.T.G., Provost, S.C., 1998. Vitamin D 3 enhances mood in healthy subjects during winter. Psychopharmacology (Berl.) 135, 319–323. doi:10.1007/s002130050517

Latulipe, C., Gatto, A., Nguyen, H.T., Miller, D.P., Quandt, S.A., Bertoni, A.G., Smith, A., Arcury, T.A., 2015. Design Considerations for Patient Portal Adoption by Low-Income, Older Adults, in: Proceedings of the 33rd Annual ACM Conference on Human Factors in Computing Systems, CHI '15. ACM, New York, NY, USA, pp. 3859–3868. doi:10.1145/2702123.2702392

Laver, K., George, S., Ratcliffe, J., Quinn, S., Whitehead, C., Davies, O., Crotty, M., 2012. Use of an interactive video gaming program compared with conventional physiotherapy for hospitalised older adults: a feasibility trial. Disabil. Rehabil. 34, 1802–1808. doi:10.3109/09638288.2012.662570

Lee, C., Myrick, R., D'Ambrosio, L.A., Coughlin, J.F., de Weck, O.L., 2013. Older Adults' Experiences with Technology: Learning from Their Voices, in: Stephanidis, C. (Ed.), HCI International 2013 - Posters' Extended Abstracts: International Conference, HCI International 2013, Las Vegas, NV, USA, July 21-26, 2013, Proceedings, Part I. Springer Berlin Heidelberg, Berlin, Heidelberg, pp. 251–255.

Lee, M.L., Dey, A.K., 2011. Reflecting on pills and phone use: supporting awareness of functional abilities for older adults, in: Proceedings of the SIGCHI Conference on Human Factors in Computing Systems. Presented at the CHI '11, ACM Press, pp. 2095–2104. doi:10.1145/1978942.1979247

Lee, Y.S., Chaysinh, S., Basapur, S., Metcalf, C.J., Mandalia, H., 2012. Active aging in community centers and ICT design implications, in: Proceedings of the Designing Interactive Systems Conference. Presented at the DIS '12, ACM Press, pp. 156–165. doi:10.1145/2317956.2317981

Lehoux, P., Gauthier, P., Williams-Jones, B., Miller, F.A., Fishman, J.R., Hivon, M., Vachon, P., 2014. Examining the ethical and social issues of health technology design through the public appraisal of prospective scenarios: a study protocol describing a multimedia-based deliberative method. Implement. Sci. 9. doi:10.1186/1748-5908-9-81

Lewin, S., Glenton, C., Oxman, A.D., 2009. Use of qualitative methods alongside randomised controlled trials of complex healthcare interventions: methodological study. BMJ 339, b3496–b3496. doi:10.1136/bmj.b3496

Ley, B., Ogonowski, C., Mu, M., Hess, J., Race, N., Randall, D., Rouncefield, M., Wulf, V., 2015. At home with users: a comparative view of Living Labs. Interact. Comput. 27, 21–35.

Li, I., Dey, A., Forlizzi, J., 2010. A stage-based model of personal informatics systems, in: Proceedings of the SIGCHI Conference on Human Factors in Computing Systems. Presented at the CHI '10, ACM Press, pp. 557–566. doi:10.1145/1753326.1753409

Lievens, B., Milić-Frayling, N., Lerouge, V., Pierson, J., Oleksik, G., Jones, R., Costello, J., 2010. Managing social adoption and technology adaption in longitudinal studies of mobile media applications, in: Proceedings of the 9th International Conference on Mobile and Ubiquitous Multimedia. Presented at the MUM '10, ACM Press, pp. 1–10. doi:10.1145/1899475.1899501

Lin, J.J., Mamykina, L., Lindtner, S., Delajoux, G., Strub, H.B., 2006. Fish'n'Steps: Encouraging Physical Activity with an Interactive Computer Game, in: Dourish, P., Friday, A. (Eds.), UbiComp 2006: Ubiquitous Computing. Springer Berlin Heidelberg, Berlin, Heidelberg, pp. 261–278. doi:10.1007/11853565_16

Lindsay, S., Jackson, D., Schofield, G., Olivier, P., 2012. Engaging older people using participatory design, in: Proceedings of the SIGCHI Conference on Human Factors in Computing Systems. Presented at the CHI 12, ACM Press, pp. 1199–1208. doi:10.1145/2207676.2208570

Locke, E.A., Latham, G.P., 2002. Building a practically useful theory of goal setting and task motivation. A 35-year odyssey. Am. Psychol. 57, 705–717.

Lord, S.R., Menz, H.B., Tiedemann, A., 2003. A physiological profile approach to falls risk assessment and prevention. Phys. Ther. 83, 237–252.

Lord, S.R., Tiedemann, A., Chapman, K., Munro, B., Murray, S.M., Gerontology, M., Ther, G.R., Sherrington, C., 2005. The effect of an individualized fall prevention program on fall risk and falls in older people: a randomized, controlled trial. J. Am. Geriatr. Soc. 53, 1296–1304. doi:10.1111/j.1532-5415.2005.53425.x

Lorenz, A., Oppermann, R., 2009. Mobile health monitoring for the elderly: Designing for diversity. Pervasive Mob. Comput. 5, 478–495. doi:10.1016/j.pmcj.2008.09.010

Lorenzen-Huber, L., Boutain, M., Camp, L.J., Shankar, K., Connelly, K.H., 2011. Privacy, Technology, and Aging: A Proposed Framework. Ageing Int. 36, 232–252. doi:10.1007/s12126-010-9083-y

Mansi, S., Milosavljevic, S., Tumilty, S., Hendrick, P., Higgs, C., Baxter, D.G., 2015. Investigating the effect of a 3-month workplace-based pedometer-driven walking programme on health-related quality of life in meat processing workers: a feasibility study within a randomized controlled trial. BMC Public Health 15. doi:10.1186/s12889-015-1736-z

Marcotte, L., Kirtane, J., Lynn, J., McKethan, A., 2015. Integrating Health Information Technology to Achieve Seamless Care Transitions: J. Patient Saf. 11, 185–190. doi:10.1097/PTS.0000000000000077

Marquié, J.C., Jourdan-Boddaert, L., Huet, N., 2002. Do older adults underestimate their actual computer knowledge? Behav. Inf. Technol. 21, 273–280. doi:10.1080/0144929021000020998

Marston, H.R., Woodbury, A., Gschwind, Y.J., Kroll, M., Fink, D., Eichberg, S., Kreiner, K., Ejupi, A., Annegarn, J., De Rosario, H., Wienholtz, A., Wieching, R., Delbaere, K., 2015. The Design of a Purpose-Built Exergame for Fall Prediction and Prevention for Older Adults. Eur. Rev. Aging Phys. Act. in press.

Martin, K.A., Bowen, D.J., Dunbar-Jacob, J., Perri, M.G., 2000. Who Will Adhere? Key Issues in the Study and Prediction of Adherence in Randomized Controlled Trials. Control. Clin. Trials 21, S195–S199. doi:10.1016/S0197-2456(00)00078-7

Mason, K., 2016. Health technologies: are older people interested?, 2020Health - making health personal. 2020Health.org, London.

Mayring, P., 2003. Qualitative Inhaltsanalyse. Grundlagen und Techniken. Beltz Dtsch. Stud. Verl. 6.

Mayring, P., 2000. Qualitative Content Analysis. Forum Qual. Sozialforschung Forum Qual. Soc. Res. 1.

McArthur, L.H., Raedeke, T.D., 2009. Race and sex differences in college student physical activity correlates. Am. J. Health Behav. 33, 80–90.

McMurdo, M.E.T., Sugden, J., Argo, I., Boyle, P., Johnston, D.W., Sniehotta, F.F., Donnan, P.T., 2010. Do Pedometers Increase Physical Activity in Sedentary Older Women? A Randomized Controlled Trial: PEDOMETERS AND PHYSICAL ACTIVITY. J. Am. Geriatr. Soc. 58, 2099–2106. doi:10.1111/j.1532-5415.2010.03127.x

Mendoza, A., 2009. 9780982261101_Content. CopaVin.

Mercer, K., Giangregorio, L., Schneider, E., Chilana, P., Li, M., Grindrod, K., 2016. Acceptance of Commercially Available Wearable Activity Trackers Among Adults Aged Over 50 and With Chronic Illness: A Mixed-Methods Evaluation. JMIR MHealth UHealth 4, e7. doi:10.2196/mhealth.4225

Merleau-Ponty, M.M., 2002. Phenomenology of perception. Routledge Classics (London, New York).

Mertens, D.M., Hesse-Biber, S., 2012. Triangulation and Mixed Methods Research: Provocative Positions. J. Mix. Methods Res. 6, 75–79. doi:10.1177/1558689812437100

Michaels, C., Carello, C., 1981. Direct perceptions. Prentice Hall.

Mihailidis, A., Cockburn, A., Longley, C., Boger, J., 2008. The Acceptability of Home Monitoring Technology Among Community-Dwelling Older Adults and Baby Boomers. Assist. Technol. 20, 1–12. doi:10.1080/10400435.2008.10131927

Miller, K.J., Adair, B.S., Pearce, A.J., Said, C.M., Ozanne, E., Morris, M.M., 2014. Effectiveness and feasibility of virtual reality and gaming system use at home by older adults for enabling physical activity to improve health-related domains: a systematic review. Age Ageing 43, 188–195. doi:10.1093/ageing/aft194

Miller, L.M.S., Bell, R.A., 2012. Online Health Information Seeking: The Influence of Age, Information Trustworthiness, and Search Challenges. J. Aging Health 24, 525–541. doi:10.1177/0898264311428167

Minnick, J., 2016. Web Design with HTML & CSS3: Comprehensive. Cengage Learning.

Morgan, D.L., 1998. Practical Strategies for Combining Qualitative and Quantitative Methods: Applications to Health Research. Qual. Health Res. 8, 362–376. doi:10.1177/104973239800800307

Morris, M.G., Venkatesh, V., 2000. AGE DIFFERENCES IN TECHNOLOGY ADOPTION DECISIONS: IMPLICATIONS FOR A CHANGING WORK FORCE. Pers. Psychol. 53, 375–403. doi:10.1111/j.1744-6570.2000.tb00206.x

Morris, S.S., Flores, R., Olinto, P., Medina, J.M., 2004. Monetary incentives in primary health care and effects on use and coverage of preventive health care interventions in rural Honduras: cluster randomised trial. The Lancet 364, 2030–2037. doi:10.1016/S0140-6736(04)17515-6

Mueller, F. "Floyd," Edge, D., Vetere, F., Gibbs, M.R., Agamanolis, S., Bongers, B., Sheridan, J.G., 2011. Designing sports: a framework for exertion games, in: Proceedings of the SIGCHI Conference on Human Factors in Computing Systems. Presented at the CHI '11, ACM Press, pp. 2651–2660. doi:10.1145/1978942.1979330

Mueller, F. "Floyd," Stevens, G., Thorogood, A., O'Brien, S., Wulf, V., 2007. Sports over a Distance. Pers. Ubiquitous Comput. 11, 633–645. doi:10.1007/s00779-006-0133-0

Müller, C., Hornung, D., Hamm, T., Wulf, V., 2015a. Practice-based Design of a Neighborhood Portal: Focusing on Elderly Tenants in a City Quarter Living Lab, in: Proceedings of the 33rd Annual ACM Conference on Human Factors in Computing Systems. Presented at the CHI2015, ACM Press, Seoul, pp. 2295–2304.

Müller, C., Hornung, D., Hamm, T., Wulf, V., 2015b. Measures and Tools for Supporting ICT Appropriation by Elderly and Non Tech-Savvy Persons in a Long-Term Perspective, in: Boulus-Rødje, N., Ellingsen, G., Bratteteig, T., Aanestad, M., Bjørn, P. (Eds.), ECSCW 2015: Proceedings of the 14th European Conference on Computer Supported Cooperative Work, 19-23 September 2015, Oslo, Norway. Springer International Publishing, Cham, pp. 263–281. doi:10.1007/978-3-319-20499-4_14

Müller, C., Schorch, M., Wieching, R., 2014. PraxLabs as a Setting for Participatory Technology Research and Design in the Field of HRI and Demography, in: Proceedings of the Workshop "Socially Assistive Robots for the Aging Population: Are We Trapped in Stereotypes?" Presented at the Human Robot Interaction Conference, Bielefeld.

Muller, M.J., 2003. The Human-computer Interaction Handbook, in: Jacko, J.A., Sears, A. (Eds.), . L. Erlbaum Associates Inc., Hillsdale, NJ, USA, pp. 1051–1068.

Muller, M.J., Kuhn, S., 1993. Participatory design. Commun. ACM - Spec. Issue Particip. Des. 36, 24–28. doi:10.1145/153571.255960

Mulvenna, M., Martin, S., McDade, D., Beamish, E., De Oliveira, A., Kivilehto, A., 2011. TRAIL Living Labs Survey 2011: A survey of the ENOLL living labs. University of Ulster.

Murray, G., 2003. Seasonality and circadian phase delay: prospective evidence that winter lowering of mood is associated with a shift towards Eveningness. J. Affect. Disord. 76, 15–22. doi:10.1016/S0165-0327(02)00059-9

Murtagh, M.J., Thomson, R.G., May, C.R., Rapley, T., Heaven, B.R., Graham, R.H., Kaner, E.F., Stobbart, L., Eccles, M.P., 2007. Qualitative methods in a randomised controlled trial: the role of an integrated qualitative process evaluation in providing evidence to discontinue the intervention in one arm of a trial of a decision support tool. Qual. Saf. Health Care 16, 224–229. doi:10.1136/qshc.2006.018499

Nawaz, A., Helbostad, J.L., Skj\a eret, N., Vereijken, B., Bourke, A., Dahl, Y., Mellone, S., 2014. Designing Smart Home Technology for Fall Prevention in Older People, in: HCI International 2014-Posters' Extended Abstracts. Springer, pp. 485–490.

Newell, A.F., Dickinson, A., Smith, M.J., Gregor, P., 2006. Designing a portal for older users: A case study of an industrial/academic collaboration. ACM Trans. Comput.-Hum. Interact. 13, 347–375. doi:10.1145/1183456.1183459

Neyens, J.C.L., Dijcks, B.P.J., Twisk, J., Schols, J.M.G.A., van Haastregt, J.C.M., van den Heuvel, W.J.A., de Witte, L.P., 2008. A multifactorial intervention for the prevention of falls in psychogeriatric nursing home patients, a randomised controlled trial (RCT). Age Ageing 38, 194–199. doi:10.1093/ageing/afn297

Nitz, J.C., Kuys, S., Isles, R., Fu, S., 2010. Is the Wii Fit™ a new-generation tool for improving balance, health and well-being? A pilot study. Climacteric 13, 487–491. doi:10.3109/13697130903395193

O'Cathain, A., Thomas, K.J., Drabble, S.J., Rudolph, A., Hewison, J., 2013. What can qualitative research do for randomised controlled trials? A systematic mapping review. BMJ Open 3, e002889. doi:10.1136/bmjopen-2013-002889

Ogonowski, C., Aal, K., Vaziri, D., von Rekowski, T., Randall, D., Schreiber, D., Wieching, R., Wulf, V., 2016. ICT-based fall prevention system for older adults: qualitative results from a long-term field study. ACM Trans Comput-Hum Interact 23.

Ogonowski, C., Aal, K., Vaziri, D., von Rekowski, T., Wieching, R., Wulf, V., 2014. A Fall Prevention Exergame for Community-Dwelling Older Adults: Results from a Long-Term Living Lab Study, under review. Presented at the CHI2014.

Ogonowski, C., Ley, B., Hess, J., Wan, L., Wulf, V., 2013. Designing for the living room: long-term user involvement in a living lab, in: Proceedings of the SIGCHI Conference on Human Factors in Computing Systems. Presented at the CHI '13, ACM Press, pp. 1539–1548. doi:10.1145/2470654.2466205

Olson, C.M., 2016. Behavioral Nutrition Interventions Using e- and m-Health Communication Technologies: A Narrative Review. Annu. Rev. Nutr. 36, 647–664. doi:10.1146/annurev-nutr-071715-050815

Orruño, E., Gagnon, M.P., Asua, J., Abdeljelil, A.B., 2011. Evaluation of teledermatology adoption by health-care professionals using a modified Technology Acceptance Model. J. Telemed. Telecare 17, 303–307. doi:10.1258/jtt.2011.101101

Pallot, M., 2009. Engaging Users into Research and Innovation: The Living Lab Approach as a User Centred Open Innovation Ecosystem. Webergence Blog.

Panek, P., Rauhala, M., Zagler, W.L., 2007. Towards a living lab for old people and their carers as co-creators of ambient assisted living (aal) technologies and applications. Presented at the Challenges for Assistive Technology, proc of the 9th Europ Conf for the Advancement of Assistive Technology in Europe (AAATE), pp. 821–825.

Papanek, V.J., 1983. Design for Human Scale. Van Nostrand Reinhold.

Phillips, E.M., Schneider, J.C., Mercer, G.R., 2004. Motivating elders to initiate and maintain exercise. Arch. Phys. Med. Rehabil. 85, 52–57.

Phiriyapokanon, T., 2011. Is a big button interface enough for elderly users. User Interface Guidel. Elder. Users Swed. Mälardalen Univ. Thesis Master Comput. Eng.

Pichierri, G., Murer, K., de Bruin, E.D., 2012. A cognitive-motor intervention using a dance video game to enhance foot placement accuracy and gait under dual task conditions in older adults: a randomized controlled trial. BMC Geriatr. 12, 74. doi:10.1186/1471-2318-12-74

Pilemalm, S., Timpka, T., 2008. Third generation participatory design in health informatics—Making user participation applicable to large-scale information system projects. J. Biomed. Inform. 41, 327–339. doi:10.1016/j.jbi.2007.09.004

Pipek, V., Wulf, V., 2009. Infrastructuring: Toward an Integrated Perspective on the Design and Use of Information Technology. J. Assoc. Inf. Syst. 10, Article 1. doi:http://aisel.aisnet.org/jais/vol10/iss5/1

Plotnikoff, R.C., Costigan, S.A., Karunamuni, N., Lubans, D.R., 2013. Social cognitive theories used to explain physical activity behavior in adolescents: A systematic review and meta-analysis. Prev. Med. 56, 245–253. doi:10.1016/j.ypmed.2013.01.013

Pluchino, A., Lee, S.Y., Asfour, S., Roos, B.A., Signorile, J.F., 2012. Pilot Study Comparing Changes in Postural Control After Training Using a Video Game Balance Board Program and 2 Standard Activity-Based Balance Intervention Programs. Arch. Phys. Med. Rehabil. 93, 1138–1146. doi:10.1016/j.apmr.2012.01.023

Procter, R., Greenhalgh, T., Wherton, J., Sugarhood, P., Rouncefield, M., Hinder, S., 2014. The Day-to-Day Co-Production of Ageing in Place. Comput. Support. Coop. Work CSCW 23, 245–267. doi:10.1007/s10606-014-9202-5

Proctor, E.K., Landsverk, J., Aarons, G., Chambers, D., Glisson, C., Mittman, B., 2009. Implementation Research in Mental Health Services: an Emerging Science with Conceptual, Methodological, and Training challenges. Adm. Policy Ment. Health Ment. Health Serv. Res. 36, 24–34. doi:10.1007/s10488-008-0197-4

Publications Office of the European Union, 2012. Redesigning Health in Europe for 2020, eHealth Task Force Report. European Commission.

Puts, M.T.E., Toubasi, S., Andrew, M.K., Ashe, M.C., Ploeg, J., Atkinson, E., Ayala, A.P., Roy, A., Rodríguez Monforte, M., Bergman, H., McGilton, K., 2017. Interventions to prevent or reduce the level of frailty in community-dwelling older adults: a scoping review of the literature and international policies. Age Ageing. doi:10.1093/ageing/afw247

Raedeke, T.D., 2007. The relationship between enjoyment and affective responses to exercise. J. Appl. Sport Psychol. 19, 105–115.

Rapp, K., Lamb, S.E., Büchele, G., Lall, R., Lindemann, U., Becker, C., 2008. Prevention of Falls in Nursing Homes: Subgroup Analyses of a Randomized Fall Prevention Trial: PREVENTION OF FALLS IN NURSING HOMES. J. Am. Geriatr. Soc. 56, 1092–1097. doi:10.1111/j.1532-5415.2008.01739.x

Raptis, D., Tselios, N., Kjeldskov, J., Skov, M.B., 2013. Does size matter?: investigating the impact of mobile phone screen size on users' perceived usability, effectiveness and efficiency., in: Proceedings of the 15th International Conference on Human-Computer Interaction with Mobile Devices and Services. ACM, pp. 127–136.

Reboussin, B.A., Rejeski, W.J., Martin, K.A., Callahan, K., Dunn, A.L., King, A.C., Sallis, J.F., 2000. Correlates of satisfaction with body function and body appearance in middle- and older aged adults: The activity counseling trial (ACT). Psychol. Health 15, 239–254. doi:10.1080/08870440008400304

Rechel, B., Grundy, E., Robine, J.M., Cylus, J., Mackenbach, J.P., Knai, C., McKee, M., 2013. Ageing in the European Union. The Lancet 381, 1312–1322. doi:10.1016/S0140-6736(12)62087-X

Regterschot, G.R.H., Folkersma, M., Zhang, W., Baldus, H., Stevens, M., Zijlstra, W., 2014. Sensitivity of sensor-based sit-to-stand peak power to the effects of training leg strength, leg power and balance in older adults. Gait Posture 39, 303–307. doi:10.1016/j.gaitpost.2013.07.122

Regterschot, G.R.H., Zhang, W., Baldus, H., Stevens, M., Zijlstra, W., 2015. Sensor-based monitoring of sit-to-stand performance is indicative of objective and self-reported aspects of functional status in older adults. Gait Posture 41, 935–940. doi:10.1016/j.gaitpost.2015.03.350

Rimmer, J.H., Riley, B., Wang, E., Rauworth, A., Jurkowski, J., 2004. Physical activity participation among persons with disabilities. Am. J. Prev. Med. 26, 419–425. doi:10.1016/j.amepre.2004.02.002

Robertson, M.C., Devlin, N., Gardner, M.M., Campbell, A.J., 2001. Effectiveness and economic evaluation of a nurse delivered home exercise programme to prevent falls. 1: Randomised controlled trial. BMJ 322, 697–701.

Robinson, L., Brittain, K., Lindsay, S., Jackson, D., Olivier, P., 2009. Keeping In Touch Everyday (KITE) project: developing assistive technologies with people with dementia and their carers to promote independence. Int. Psychogeriatr. 21, 494. doi:10.1017/S1041610209008448

Rogers, W.A., Fisk, A.D., Mead, S.E., Walker, N., Cabrera, E.F., 1996. Training Older Adults to Use Automatic Teller Machines. Hum. Factors J. Hum. Factors Ergon. Soc. 38, 425–433. doi:10.1518/001872096778701935

Rohde, M., Stevens, G., Brödner, P., Wulf, V., 2009. Towards a paradigmatic shift in IS: designing for social practice, in: Proceedings of the 4th International Conference on Design Science Research in Information Systems and Technology. Presented at the DESRIST '09, ACM Press, p. 1. doi:10.1145/1555619.1555639

Rooksby, J., Rost, M., Morrison, A., Chalmers, M.C., 2014. Personal tracking as lived informatics, in: Proceedings of the SIGCHI Conference on Human Factors in Computing Systems. Presented at the CHI '14, ACM Press, pp. 1163–1172. doi:10.1145/2556288.2557039

Scarapicchia, T.M.F., Amireault, S., Faulkner, G., Sabiston, C.M., 2017. Social support and physical activity participation among healthy adults: a systematic review of prospective studies. Int. Rev. Sport Exerc. Psychol. 10, 50–83. doi:10.1080/1750984X.2016.1183222

Schaefbauer, C.L., Khan, D.U., Le, A., Sczechowski, G., Siek, K.A., 2015. Snack Buddy: Supporting Healthy Snacking in Low Socioeconomic Status Families, in: Proceedings of the 18th ACM Conference on Computer Supported Cooperative

Work & Social Computing, CSCW '15. ACM, New York, NY, USA, pp. 1045–1057. doi:10.1145/2675133.2675180

Schaffers, H., Sällström, A., Pallot, M., Hernandez-Munoz, J.M., Santoro, R., Trousee, B., 2011. Integrating Living Labs with Future Internet experimental platforms for co-creating services within Smart Cities, in: 17th International Conference on Concurrent Enterprising (ICE), 2011: 20 - 22 June 2011, Aachen, Germany. Presented at the 2011 17th International Conference on Concurrent Enterprising (ICE), IEEE, Aachen.

Schlomann, A., von Storch, K., Rasche, P., Rietz, C., 2016. Means of Motivation or of Stress? The Use of Fitness Trackers for Self-Monitoring by Older Adults. HeilberufeScience 7, 111–116. doi:10.1007/s16024-016-0275-6

Schoene, D., Lord, S.R., Delbaere, K., Severino, C., Davies, T.A., Smith, S.T., 2013. A Randomized Controlled Pilot Study of Home-Based Step Training in Older People Using Videogame Technology. PLoS ONE 8, e57734. doi:10.1371/journal.pone.0057734

Schoene, D., Lord, S.R., Verhoef, P., Smith, S.T., 2011. A novel video game–based device for measuring stepping performance and fall risk in older people. Arch. Phys. Med. Rehabil. 92, 947–953.

Schoene, D., Valenzuela, T., Toson, B., Delbaere, K., Severino, C., Garcia, J., Davies, T.A., Russell, F., Smith, S.T., Lord, S.R., 2015. Interactive Cognitive-Motor Step Training Improves Cognitive Risk Factors of Falling in Older Adults – A Randomized Controlled Trial. PLOS ONE 10, e0145161. doi:10.1371/journal.pone.0145161

Schorch, M., Wan, L., Randall, D.W., Wulf, V., 2016. Designing for Those Who Are Overlooked: Insider Perspectives on Care Practices and Cooperative Work of Elderly Informal Caregivers, in: Proceedings of the 19th ACM Conference on Computer-Supported Cooperative Work & Social Computing, CSCW '16. ACM, New York, NY, USA, pp. 787–799. doi:10.1145/2818048.2819999

Schulz, R., Wahl, H.-W., Matthews, J.T., De Vito Dabbs, A., Beach, S.R., Czaja, S.J., 2015. Advancing the Aging and Technology Agenda in Gerontology. The Gerontologist 55, 724–734. doi:10.1093/geront/gnu071

Schutzer, K.A., 2004. Barriers and motivations to exercise in older adults. Prev. Med. 39, 1056–1061. doi:10.1016/j.ypmed.2004.04.003

Sellen, A.J., Whittaker, S., 2010. Beyond total capture: a constructive critique of lifelogging. Commun. ACM 53, 70. doi:10.1145/1735223.1735243

Selwyn, N., 2004. The information aged: A qualitative study of older adults' use of information and communications technology. J. Aging Stud. 18, 369–384. doi:10.1016/j.jaging.2004.06.008

Sheikh, J.A., Abbas, A., 2015. The Invisible User: Women in DUXU, in: Design, User Experience, and Usability: Users and Interactions. Springer, pp. 243–251.

Sherrington, C., Tiedemann, A., Fairhall, N., Close, J.C.T., Lord, S.R., 2011. Exercise to prevent falls in older adults: an updated meta-analysis and best practice recommendations. New South Wales Public Health Bull. 22, 78. doi:10.1071/NB10056

Sherrington, C., Whitney, J.C., Lord, S.R., Herbert, R.D., Cumming, R.G., Close, J.C., 2008. Effective exercise for the prevention of falls: a systematic review and meta-analysis. J. Am. Geriatr. Soc. 56, 2234–2243.

Siek, K.A., Ross, S.E., Khan, D.U., Haverhals, L.M., Cali, S.R., Meyers, J., 2010. Colorado Care Tablet: The design of an interoperable Personal Health Application to help older adults with multimorbidity manage their medications. J. Biomed. Inform. 43, S22–S26. doi:10.1016/j.jbi.2010.05.007

Simek, E.M., McPhate, L., Haines, T.P., 2012. Adherence to and efficacy of home exercise programs to prevent falls: A systematic review and meta-analysis of the impact of exercise program characteristics. Prev. Med. 55, 262–275. doi:10.1016/j.ypmed.2012.07.007

Singh, D.K.A., Rajaratnam, B.S., Palaniswamy, V., Pearson, H., Raman, V.P., Bong, P.S., 2012. Participating in a virtual reality balance exercise program can reduce risk and fear of falls. Maturitas 73, 239–243. doi:10.1016/j.maturitas.2012.07.011

Smeddinck, J., Gerling, K.M., Tiemkeo, S., 2013. Visual complexity, player experience, performance and physical exertion in motion-based games for older adults, in: Proceedings of the 15th International ACM SIGACCESS Conference on Computers and Accessibility. Presented at the ASSETS '13, ACM Press, pp. 1–8. doi:10.1145/2513383.2517029

Smith, K.A., Dennis, M., Masthoff, J., 2016. Personalizing Reminders to Personality for Melanoma Self-checking, in: Proceedings of the 2016 Conference on User Modeling Adaptation and Personalization, UMAP '16. ACM, New York, NY, USA, pp. 85–93. doi:10.1145/2930238.2930254

Smith, S.T., Sherrington, C., Studenski, S., Schoene, D., Lord, S.R., 2011. A novel Dance Dance Revolution (DDR) system for in-home training of stepping ability: basic parameters of system use by older adults. Br. J. Sports Med. 45, 441–445. doi:10.1136/bjsm.2009.066845

Society, A.G., Society, G., Of, A.A., On Falls Prevention, O.S.P., 2001. Guideline for the Prevention of Falls in Older Persons. J. Am. Geriatr. Soc. 49, 664–672. doi:10.1046/j.1532-5415.2001.49115.x

Sørensen, K., Van den Broucke, S., Fullam, J., Doyle, G., Pelikan, J., Slonska, Z., Brand, H., (HLS-EU) Consortium Health Literacy Project European, 2012. Health literacy and public health: A systematic review and integration of definitions and models. BMC Public Health 12, 80. doi:10.1186/1471-2458-12-80

Ståhlbröst, A., 2004. Exploring the Testbed field. Presented at the 27th Information Systems Research Seminars in Scandinavia, IRIS.

Steele, R., Lo, A., Secombe, C., Wong, Y.K., 2009. Elderly persons' perception and acceptance of using wireless sensor networks to assist healthcare. Int. J. Med. Inf. 78, 788–801. doi:10.1016/j.ijmedinf.2009.08.001

Steinberg, M., Cartwright, C., Peel, N., Williams, G., 2000. A sustainable programme to prevent falls and near falls in community dwelling older people: results of a randomised trial. J. Epidemiol. Community Health 54, 227–232.

Stellefson, M., Alber, J.M., Wang, M.Q., Eddy, J.M., Chaney, B.H., Chaney, J.D., 2015. Use of Health Information and Communication Technologies to Promote Health and Manage Behavioral Risk Factors Associated With Chronic Disease: Applications in the Field of Health Education. Am. J. Health Educ. 46, 185–191. doi:10.1080/19325037.2015.1043064

Stetler, C.B., Legro, M.W., Wallace, C.M., Bowman, C., Guihan, M., Hagedorn, H., Kimmel, B., Sharp, N.D., Smith, J.L., 2006. The Role of Formative Evaluation in Implementation Research and the QUERI Experience. J. Gen. Intern. Med. 21, S1–S8. doi:10.1111/j.1525-1497.2006.00355.x

Stevens, J.A., Corso, P.S., Finkelstein, E.A., Miller, T.R., 2006. The costs of fatal and non-fatal falls among older adults. Inj. Prev. 12, 290–295. doi:10.1136/ip.2005.011015

Stevens, M., Holman, C.D., Bennett, N., de Klerk, N., 2001. Preventing falls in older people: outcome evaluation of a randomized controlled trial. J. Am. Geriatr. Soc. 49, 1448–1455.

Studenski, S., Perera, S., Hile, E., Keller, V., Spadola-Bogard, J., Garcia, J., 2010. Interactive video dance games for healthy older adults. J. Nutr. Health Aging 14, 850–852. doi:10.1007/s12603-010-0119-5

Tange, H., Genderen, S., Van der Weegen, S., Moser, A., Plasqui, G., 2012. A pilot with Exergames in Elderly Homes, in: 23rd International Conference of the European Federation for Medical Informatics: User Centred Networked Health Care.

Teddlie, C., Tashakkori, A., 2006. A general typology of research designs featuring mixed methods. Res. Sch. 13, 12–28.

Thomas, J.G., Bond, D.S., 2014. Review of Innovations in Digital Health Technology to Promote Weight Control. Curr. Diab. Rep. 14. doi:10.1007/s11892-014-0485-1

Thompson, H.J., Demiris, G., Rue, T., Shatil, E., Wilamowska, K., Zaslavsky, O., Reeder, B., 2011. A Holistic Approach to Assess Older Adults' Wellness Using e-Health Technologies. Telemed. E-Health 17, 794–800. doi:10.1089/tmj.2011.0059

Thraen-Borowski, K.M., Trentham-Dietz, A., Edwards, D.F., Koltyn, K.F., Colbert, L.H., 2013. Dose–response relationships between physical activity, social participation, and health-related quality of life in colorectal cancer survivors. J. Cancer Surviv. 7, 369–378. doi:10.1007/s11764-013-0277-7

Tiedemann, A., Shimada, H., Sherrington, C., Murray, S., Lord, S., 2008. The comparative ability of eight functional mobility tests for predicting falls in community-dwelling older people. Age Ageing 37, 430–435. doi:10.1093/ageing/afn100

Tobiasson, H., Hedman, A., Sundblad, Y., 2012. Design space and opportunities for physical movement participation in everyday life, in: Proceedings of the 24th

Australian Computer-Human Interaction Conference. Presented at the OzCHI '12, ACM Press, pp. 607–615. doi:10.1145/2414536.2414628

Topolovec-Vranic, J., Mansoor, Y., Ennis, N., Lightfoot, D., 2015. Technology-adaptable interventions for treating depression in adults with cognitive impairments: protocol for a systematic review. Syst. Rev. 4. doi:10.1186/s13643-015-0032-4

Torning, K., Oinas-Kukkonen, H., 2009. Persuasive system design: state of the art and future directions, in: Proceedings of the 4th International Conference on Persuasive Technology. Presented at the Persuasive '09, ACM Press, p. 1. doi:10.1145/1541948.1541989

Torun, M., van Kasteren, T., Durmaz Incel, O., Ersoy, C., 2012. Complexity versus Page Hierarchy of a GUI for Elderly Homecare Applications, in: Miesenberger, K., Karshmer, A., Penaz, P., Zagler, W. (Eds.), Computers Helping People with Special Needs. Springer Berlin Heidelberg, Berlin, Heidelberg, pp. 689–696. doi:10.1007/978-3-642-31522-0_103

Trujillo, K.M., Brougham, R.R., Walsh, D.A., 2004. Age differences in reasons for exercising. Curr. Psychol. 22, 348–367. doi:10.1007/s12144-004-1040-z

Tudor-Locke, C., 2002. Taking Steps toward Increased Physical Activity: Using Pedometers To Measure and Motivate. Pres. Counc. Phys. Fit. Sports Res. Dig.

Tudor-Locke, C., Bassett, D.R., 2004. How Many Steps/Day Are Enough?: Preliminary Pedometer Indices for Public Health. Sports Med. 34, 1–8. doi:10.2165/00007256-200434010-00001

UNHCR, 2015. UNHCR Global Appeal 2015 Update - Populations of concern to UNHCR [WWW Document]. UNHCR Glob. Appeal 2015 Update - Popul. Concern UNHCR. URL http://www.unhcr.org/5461e5ec3c.html

Uzor, S., Baillie, L., 2014. Investigating the long-term use of exergames in the home with elderly fallers, in: Proceedings of the SIGCHI Conference on Human Factors in Computing Systems. Presented at the CHI '14, ACM Press, pp. 2813–2822. doi:10.1145/2556288.2557160

Uzor, S., Baillie, L., 2013. Exploring and designing tools to enhance falls rehabilitation in the home, in: Proceedings of the SIGCHI Conference on Human Factors in Computing Systems. Presented at the CHI '13, ACM Press, pp. 1233–1242. doi:10.1145/2470654.2466159

Uzor, S., Baillie, L., Skelton, D., 2012. Senior designers: empowering seniors to design enjoyable falls rehabilitation tools, in: Proceedings of the SIGCHI Conference on Human Factors in Computing Systems. Presented at the CHI '12, ACM Press, pp. 1179–1188. doi:10.1145/2207676.2208568

Valenzuela, T., Okubo, Y., Woodbury, A., Lord, S.R., Delbaere, K., 2016. Adherence to Technology-Based Exercise Programs in Older Adults: A Systematic Review. J. Geriatr. Phys. Ther. Publish Ahead of Print.

van Gemert-Pijnen, J.E., Nijland, N., van Limburg, M., Ossebaard, H.C., Kelders, S.M., Eysenbach, G., Seydel, E.R., 2011. A Holistic Framework to Improve the Uptake and Impact of eHealth Technologies. J. Med. Internet Res. 13, e111. doi:10.2196/jmir.1672

Van Schaik, P., Bettany-Saltikov, J.A., Warren, J.G., 2002. Clinical acceptance of a low-cost portable system for postural assessment. Behav. Inf. Technol. 21, 47–57. doi:10.1080/01449290110107236

Vankipuram, M., McMahon, S., Fleury, J., 2012. ReadySteady: App for Accelerometer-based Activity Monitoring and Wellness-Motivation Feedback System for Older Adults. AMIA. Annu. Symp. Proc. 2012, 931–939.

Vaziri, D.D., Aal, K., Gschwind, Y.J., Delbaere, K., Weibert, A., Annegarn, J., de Rosario, H., Wieching, R., Randall, D., Wulf, V., 2017. Analysis of effects and usage indicators for a ICT-based fall prevention system in community dwelling older adults. Int. J. Hum.-Comput. Stud. 106, 10–25. doi:10.1016/j.ijhcs.2017.05.004

Vaziri, D.D., Aal, K., Ogonowski, C., Von Rekowski, T., Kroll, M., Marston, H.R., Poveda, R., Gschwind, Y.J., Delbaere, K., Wieching, R., Wulf, V., 2016. Exploring user experience and technology acceptance for a fall prevention system: results from a randomized clinical trial and a living lab. Eur. Rev. Aging Phys. Act. 13. doi:10.1186/s11556-016-0165-z

Verbeek, P.P., 2010. What Things Do: Philosophical Reflections on Technology, Agency, and Design. Pennsylvania State University Press.

Verwey, R., van der Weegen, S., Spreeuwenberg, M., Tange, H., van der Weijden, T., de Witte, L., 2014. A monitoring and feedback tool embedded in a counselling protocol to increase physical activity of patients with COPD or type 2 diabetes in primary care: study protocol of a three-arm cluster randomised controlled trial. BMC Fam. Pract. 15. doi:10.1186/1471-2296-15-93

Vines, J., Pritchard, G., Wright, P., Olivier, P., Brittain, K., 2015. An Age-Old Problem: Examining the Discourses of Ageing in HCI and Strategies for Future Research. ACM Trans. Comput.-Hum. Interact. 22, 1–27. doi:10.1145/2696867

Volpp, K.G., Asch, D.A., Galvin, R., Loewenstein, G., 2011. Redesigning Employee Health Incentives — Lessons from Behavioral Economics. N. Engl. J. Med. 365, 388–390. doi:10.1056/NEJMp1105966

Wan, L., Müller, C., Randall, D., Wulf, V., 2016. Design of A GPS Monitoring System for Dementia Care and its Challenges in Academia-Industry Project. ACM Trans. Comput.-Hum. Interact. 23, 1–36. doi:10.1145/2963095

Wan, L., Müller, C., Wulf, V., Randall, D.W., 2014. Addressing the subtleties in dementia care: pre-study and evaluation of a GPS monitoring system, in: Proceedings of the SIGCHI Conference on Human Factors in Computing Systems. Presented at the CHI '14, ACM Press, pp. 3987–3996. doi:10.1145/2556288.2557307

Westfall, J.M., Mold, J., Fagnan, L., 2007. Practice-based research--"Blue Highways" on the NIH roadmap. JAMA 297, 403–406. doi:10.1001/jama.297.4.403

White, H., McConnell, E., Clipp, E., Branch, L.G., Sloane, R., Pieper, C., Box, T.L., 2002. A randomized controlled trial of the psychosocial impact of providing internet training and access to older adults. Aging Ment. Health 6, 213–221. doi:10.1080/13607860220142422

White, H., McConnell, E., Clipp, E., Bynum, L., Teague, C., Navas, L., Craven, S., Halbrecht, H., 1999. Surfing the Net in Later Life: A Review of the Literature and Pilot Study of Computer Use and Quality of Life. J. Appl. Gerontol. 18, 358–378. doi:10.1177/073346489901800306

Whitehead, C.H., Wundke, R., Crotty, M., 2006. Attitudes to falls and injury prevention: what are the barriers to implementing falls prevention strategies? Clin. Rehabil. 20, 536–542. doi:10.1191/0269215506cr984oa

Wiedenbeck, S., 1999. The use of icons and labels in an end user application program: An empirical study of learning and retention. Behav. Inf. Technol. 18, 68–82. doi:10.1080/014492999119129

Wiemeyer, J., 2010. Serious Games–The challenges for computer science in sport. Int. J. Comput. Sci. Sport 9.

Wilkowska, W., Ziefle, M., 2009. Which Factors Form Older Adults' Acceptance of Mobile Information and Communication Technologies?, in: Holzinger, A., Miesenberger, K. (Eds.), HCI and Usability for E-Inclusion. Springer Berlin Heidelberg, Berlin, Heidelberg, pp. 81–101.

Williams, B., Doherty, N.L., Bender, A., Mattox, H., Tibbs, J.R., 2011. The Effect of Nintendo Wii on Balance: A Pilot Study Supporting the Use of the Wii in Occupational Therapy for the Well Elderly. Occup. Ther. Health Care 25, 131–139. doi:10.3109/07380577.2011.560627

Williams, D.M., Anderson, E.S., Winett, R.A., 2005. A review of the outcome expectancy construct in physical activity research. Ann. Behav. Med. 29, 70–79. doi:10.1207/s15324796abm2901_10

Williams, M.A., Soiza, R.L., Jenkinson, A.M., Stewart, A., 2010. Exercising with Computers in Later Life (EXCELL) - pilot and feasibility study of the acceptability of the Nintendo® WiiFit in community-dwelling fallers. BMC Res. Notes 3, 1–8. doi:10.1186/1756-0500-3-238

Wolf, M.S., Gazmararian, J.A., Baker, D.W., 2005. Health Literacy and Functional Health Status Among Older Adults. Arch. Intern. Med. 165, 1946. doi:10.1001/archinte.165.17.1946

Wulf, V., 2009. Theorien sozialer Praktiken zur Fundierung der Wirtschaftsinformatik, in: Becker, J., Krcmar, H., Niehaves, B. (Eds.), Wissenschaftstheorie und gestaltungsorientierte Wirtschaftsinformatik. Physica-Verlag HD, Heidelberg, pp. 211–224. doi:10.1007/978-3-7908-2336-3_11

Wulf, V., Moritz, E.F., Henneke, C., Al-Zubaidi, K., Stevens, G., 2004. Computer Supported Collaborative Sports: Creating Social Spaces Filled with Sports Activities, in: Rauterberg, M. (Ed.), Entertainment Computing – ICEC 2004. Springer Berlin Heidelberg, Berlin, Heidelberg, pp. 80–89. doi:10.1007/978-3-540-28643-1_11

Wulf, V., Müller, C., Pipek, V., Randall, D., Rohde, M., Stevens, G., 2015a. Practice-Based Computing: Empirically Grounded Conceptualizations Derived from Design Case Studies, in: Wulf, V., Schmidt, K., Randall, D. (Eds.), Designing Socially Embedded Technologies in the Real-World. Springer London, London, pp. 111–150. doi:10.1007/978-1-4471-6720-4_7

Wulf, V., Pipek, V., Rohde, M., Schmidt, K., Stevens, G., Randall, D., 2015b. Socio Informatics – A Practice-based Perspective. Oxford University Press.

Wulf, V., Pipek, V., Won, M., 2008. Component-based tailorability: Enabling highly flexible software applications. Int. J. Hum.-Comput. Stud. 66, 1–22. doi:10.1016/j.ijhcs.2007.08.007

Wulf, V., Rohde, M., Pipek, V., Stevens, G., 2011. Engaging with practices: design case studies as a research framework in CSCW, in: Proceedings of the ACM 2011 Conference on Computer Supported Cooperative Work. Presented at the CSCW 2011, ACM, pp. 505–512.

Wulf, V., Schmidt, K., Randall, D. (Eds.), 2015c. Designing Socially Embedded Technologies in the Real-World, Computer Supported Cooperative Work. Springer London, London. doi:10.1007/978-1-4471-6720-4

Yardley, L., Bishop, F.L., Beyer, N., Hauer, K., Kempen, G.I.J.M., Piot-Ziegler, C., Todd, C.J., Cuttelod, T., Horne, M., Lanta, K., Holt, A.R., 2006. Older People's Views of Falls-Prevention Interventions in Six European Countries. The Gerontologist 46, 650–660. doi:10.1093/geront/46.5.650

Yardley, L., Kirby, S., Ben-Shlomo, Y., Gilbert, R., Whitehead, S., Todd, C., 2008. How likely are older people to take up different falls prevention activities? Prev. Med. 47, 554–558. doi:10.1016/j.ypmed.2008.09.001

Yi, M.Y., Jackson, J.D., Park, J.S., Probst, J.C., 2006. Understanding information technology acceptance by individual professionals: Toward an integrative view. Inf. Manage. 43, 350–363. doi:10.1016/j.im.2005.08.006

Zhao, D., Lustria, M.L.A., Hendrickse, J., 2017. Systematic review of the information and communication technology features of web- and mobile-based psychoeducational interventions for depression. Patient Educ. Couns. 100, 1049–1072. doi:10.1016/j.pec.2017.01.004

Printed in the United States
By Bookmasters